Accelerated LotusScript Study Guide

Accelerated LotusScript Study Guide

William Thompson

McGraw-Hill
New York • San Francisco • Washington, D.C. • Auckland
Bogotá • Caracas • Lisbon • London • Madrid • Mexico City
Milan • Montreal • New Delhi • San Juan • Singapore
Sydney • Tokyo • Toronto

McGraw-Hill

A Division of The McGraw-Hill Companies

1 2 3 4 5 6 7 8 9 0   AGM/AGM   9 0 4 3 2 1 0 9

ISBN 0-07-134561-2

The sponsoring editors for this book were Judy Brief and Michael Sprague, and the production supervisor was Pamela Pelton. It was set in Stone by D & G Limited, LLC.

Printed and bound by Quebecor/Martinsburg.

McGraw-Hill books are available at special quantity discounts to use as premiums and sales promotions, or for use in corporate training programs. For more information, please write to Director of Special Sales, McGraw-Hill, 11 West 19th Street, New York, NY 10011. Or contact your local bookstore.

This book is printed on recycled, acid-free paper containing a minimum of 50% recycled de-inked fiber.

Dedication

For my loving wife Sharon

Acknowledgments

I would like to thank my technical reviewer Sean Keighron of Whittman-Hart, Inc., for providing his valuable input and suggestions for this book.

Additional thanks go to Judy Brief and Michael Sprague, my editors at McGraw-Hill; Alan Harris, project manager at D&G Ltd.; and all the other members of McGraw-Hill and D&G Ltd., who made this book possible.

Finally, I would like to thank my family and friends who have been pillars in my life and whose constant support and inspiration made this book possible.

Contents

8   Controlling Database Access   213

9   Session Access   237

Index   **327**

INTRODUCTION

In recent years, the explosion of Lotus Notes has increased the demand for qualified Notes professionals. The Lotus Notes Certified Professional program provides a way for an individual to be recognized as knowledgeable about Notes. Although many publications are available to help you become an effective developer or system administrator, to date no publication is available to help you pass the LotusScript in Notes for Advanced Developers Certification exam. Currently the only option available is to attend expensive Lotus Notes certification classes.

The purpose of this book is twofold. First, this book will help advanced developers understand in what areas they must concentrate their study for the exam and provides numerous examples of scripts for review. Secondly, this book logically guides less experienced LotusScript developers through the foundation and critical functional areas so that they can understand what areas they must improve on to pass the exam. Many areas of LotusScript are covered in this book. This book's intention is to give you the greatest exposure to what you need to know in order to pass the exam.

Note that this reference specifically deals with the certification exams related to Lotus Notes version 4.x. The questions in this book, as well as those found on the tests, will not reflect any enhancements made to later releases. To the best of my knowledge, Lotus does not plan to distribute any additional exams until the release of Lotus Notes 5.0. I do not anticipate those exams being available until the second or third quarter of 1999.

Lotus Notes has become a dominant, defining product in the Groupware market. Many companies are deploying Lotus Notes for all of the benefits inherent to the product. In deploying Lotus Notes, knowledgeable professionals are needed to help maintain and develop business-critical systems. Certification by the software vendor is a way to verify that a professional is truly an expert.

In studying for the exam, some people take classes or study on their own, while others rely on their knowledge of the product. In any event, this book will be a resource to help people who feel they need some additional guidance.

The Certified Lotus Professional Program

The *Certified Lotus Professional* (CLP) Program provides a way to measure the technical knowledge and skills associated with Lotus Notes and Lotus cc:Mail. Certification demonstrates that a person has achieved a professional level of expertise within a particular area. Selected competencies are tested through Sylvan Prometric exams, which require not only technical training, but also practical experience. Classes are available for obtaining structured training, but this book enables you to identify the skills required to pass the exams regardless of whether you have taken any coursework.

The LotusScript in Notes for Advanced Developers Certification exam is needed to be certified as a Principal Application Developer.

To become a CLP Principal Notes Application Developer, you must successfully pass the following exams:

- Application Development 1
- System Administration 1 or Domino Messaging Administration 4.6
- Application Development 2

Electives (one required from the following exams)

- LotusScript in Notes for Advanced Developers
- Developing Domino Applications for the Web (for Notes Developers)
- Developing LotusScript Applications for SmartSuite Using 1-2-3 '97
- Domino Web Development & Administration
- Supporting Domino and Notes

Examination Types

Lotus offers two types of testing methods: multiple choice or concurrent application. Multiple choice is the standard exam format used for all exams where the candidate is presented with a question and several possible answers. Concurrent application testing is task based, which enables the candidate to perform actual tasks within a Notes environment. For example, in the LotusScript in Notes for Advanced Developers Certification exam, candidates would be presented with an actual Notes database with scripts that must be modified based on the exam questions.

 NOTE
Only multiple-choice exam questions and information will be covered in this book. Concurrent application testing within the Notes environment will NOT be covered. For more information on concurrent-application testing, please see the Lotus Notes CLP Exam guide available from Lotus.

Testing Format

The multiple choice format at Notes examinations offers the candidate 40–60 multiple choice questions per test, each of which may have one or multiple answers. Depending on which exam you are taking, the time limit on the exam may vary. For most candidates, the time allowed will be sufficient to complete the examination.

The testing environment is computerized and follows examinations similar to Microsoft certification exams, in which the student may traverse through the examination and either answer and/or mark troublesome questions. The candidate then has the ability to go back to questions and review the answers. The Lotus Notes tests are not adaptive like Novell certification examinations. You will have a set number of questions that will quiz you on all topics associated with that particular examination.

Questions will be direct and related to certain topics or will be based upon a given scenario in which the candidate must gather pertinent information.

All examinations are closed-book format, and no external notes, calculators, or computers are allowed. A blank, scratch sheet of paper will be provided to you at the examination center. That sheet has to be turned back to the center when the examination is complete.

When the examination is completed, you will be notified immediately of your test score on-screen. A printout of the score will also be provided to you. You must pass a certain percentage of questions to achieve the passing grade, which will be revealed to you when you start the examination. Your score will then be automatically forwarded to the Lotus Education Department within five business days.

 NOTE
If you do not pass the examination, you will be required to re-register for the examination and repay the registration fee.

Study Methods

Many candidates ask what is the best method for studying for the certification examinations. The standard answer is that it depends. The best method is what makes you feel the most comfortable with the material. For some, the best method includes memorizing all available information on a particular topic. For others, it is hands on experimentation with the product itself. For other still, it is sample test questions. For most, it is a combination of all three. Hopefully, this book will address the needs of all test takers.

Cheating Policies

You should completely understand the testing rules of the facility and the repercussions of failing to abide by those rules. If an examination is passed by questionable means and it is indeed proven by the testing center along with Lotus, the test will be nullified, and the candidate will not be permitted to take another certification examination for six months.

Registering for Examinations

At present, Lotus Certification examinations are offered by Sylvan Prometric Testing Centers (which may be found worldwide) and CATGlobal, which is currently rolling out testing centers world wide. For information about testing centers or to register for a Lotus examination, call one of the following numbers:

Sylvan Prometric Testing Centers

800–74–LOTUS (800–745–6887)
or
612–896–7000

You can also register via the internet at www.slspro.com.
When calling the Sylvan Prometric testing center, you will need the following items:

- Name
- Social Security number (this will be your Sylvan Prometric ID number)
- Mailing address and phone number
- Company name
- Name and code number of the examination you will be taking
- Date you wish to take the examination
- Method of payment (Credit card, Money order, or Check)

The testing center will instruct you on the location of the testing center and brief you on the cancellation policy.

 NOTE
If you leave your current employer, your certification will follow you to your next employer.

CATGlobal

You can also register online for an exam at www.catglopal.com. At the CatGlobal Web site, you will be asked to enter some personal information as well as billing information. You then will be given an ID for use at the Web site.

Other Contacts

Other information on Lotus Education may be obtained from Lotus' Web site at www.lotus.com or by calling the Lotus Education Helpline at 800-346-6409.

CHAPTER 1

LotusScript Basics

To begin writing scripts, you must understand where you can use LotusScript and when it makes sense to use LotusScript versus the formula language. As this chapter points out, it is not always prudent to use LotusScript when the formula language can perform similar functions with less code. You must also understand a little about event-driven languages. This chapter will discuss the events that are available to you and when they occur. As you learn about these events, be sure to investigate them within an application. This chapter also introduces the Lotus Notes Interactive Debugger. As you write and test your scripts, the debugger will become an invaluable tool in helping you debug your code.

Chapter Objectives

The objectives of this chapter are listed here:

- Understanding the differences between LotusScript and the formula language
- Understanding which objects (design elements) are codable
- Understanding events and when they occur within a Notes database

- Knowing when to use or not use LotusScript
- Learning how to use the interactive debugger

LotusScript and Lotus Notes

LotusScript is a full, object-oriented programming language. Its interface to Notes is through predefined object classes. Notes oversees the compilation and loading of user scripts and automatically includes the Notes class definitions. LotusScript is one of three distinct languages used by Lotus Notes if we do not count the full text search language and the language generated by Simple Actions. The other two languages used by Notes are the formula language (@Functions) and the command language (@Commands).

Even though the core language of LotusScript is the same across different products, each Lotus product that uses LotusScript includes product-specific, object-oriented extensions. This inclusion enables cross-product automation across multiple operating systems. For example, a spreadsheet might have cells that are manipulated by cell objects, whereas Notes documents do not have cells but instead have fields that would be manipulated by item objects.

As you move along in your development of Notes applications, it becomes apparent that the formula and command languages fall short in some areas which are necessary to create sophisticated applications. You must master the function and command languages, as well as LotusScript, to create sophisticated applications.

When to Use LotusScript

The decision to use one language instead of another depends on the following items:

- Whether the interface enables you to use either one or the other. For example, in a field in Notes, the Entering event will only enable you to use LotusScript. Note that the formula and Simple action(s) radio buttons are grayed out, as seen in Figure 1.1.
- The sophistication of the task you want to accomplish. A good example is looping routines. The formula language does not handle looping routines, whereas LotusScript does.

Figure 1.1   A script Interface

- Whether the task can be or is more efficiently performed using one language instead of another. In many instances, one line of code using the @function language can perform the same operation as multiple lines of code using LotusScript. Compare the following code examples

```
@Function Language:
@SetField("UC"; @UpperCase(UC))
```

This line of code will convert all of the values in the UC field to upper case by using one line of code. If the UC field is a multi-value field, achieving the equivalent result using LotusScript will require you to write code similar to the following example:

```
LotusScript:
Dim Workspace As New NotesUIWorkSpace
Dim UIDoc As NotesUIDocument
Set UIDoc=WorkSpace.CurrentDocument
Dim Doc As NotesDocument
Set Doc = UIDoc.Document
GetList = Doc.UC
GetCount% = Ubound (GetList) + 1
Dim PutBack() As String
Recim PutBack(GetCount%)
ArrayCounter%=0
Forall Element In GetList
     PutBack(ArrayCounter%) = Ucase(Element)
     ArrayCounter%=ArrayCounter%+1
End Forall
Dim Item As NotesItem
Set Item = Doc.ReplaceItemValue("UC", PutBack)
```

- And lastly, personal preference. Some people are more comfortable using one language instead of another, or they typically like to perform certain operations using one language instead of another. This is not to say that one language is better than the other, but each language has its place.

Script Versus @Functions

Table 1.1 outlines the general advantages and disadvantages of each language.

Table 1.1

Language	Function
@Function Advantages	■ Best for working within an object that a user is currently processing or interacting with, such as
	1. Field default, input translation, and input validation formulas
	2. View selection or view column formulas
	■ More robust text manipulation functions
	■ Most functions transparently handle multi-value data lists
	■ More complete control over menu commands using @Command
	■ Can access external data sources and call .DLLs using the @Db functions with little coding
	■ Faster execution that LotusScript, all things being equal
	■ Less code-writing to accomplish an equivalent task
@Function Disadvantages	■ No looping structures (If/Then, For/Next, Do/While)
	■ Cannot access many database and document properties, such as the ACL or Item properties
	■ Limited debugging capability (only simple syntax checking)

Language	Function
LotusScript Advantages	■ Better for processing existing objects, usually through agents, such as changing values in one document based on values in found in another
	■ Is extensible by using LSX object libraries. For example, the ODBC object classes are added using the *LSXODBC.DLL file.
	■ LotusScript also has the following capacities:
	1. Interacting with fields in a form and saving the data to a different Notes document or to an external database.
	2. Interacting with data and embedded objects using ODBC and OLE2.
	3. Calling operating system APIs and C libraries.
	4. Reading and writing to text fields.
	■ Interactive debugging facility
LotusScript Disadvantages	■ More difficult to program an application
	■ Must write complicated control structures to manipulate multi-value fields
	■ Not all LotusScript capability is implemented in Notes.

LotusScript and Security

The Server document in the Public Address Book determines whether you can run restricted and/or unrestricted agents. Users or groups allowed to run unrestricted agents have full access to the server's system and can manipulate system time, file I/O, and operating system commands. Users or groups allowed to run restricted agents have access to a subset of LotusScript properties and methods, which enables limited access to the server's system. Restricted and Unrestricted Notes Data Access is shown in Table 1.2.

Table 1.2   Restricted and Unrestricted Notes Data Access

Script Executed	Restricted Notes Data Access	Unrestricted Notes Data Access
Locally from menu or scheduled agent	■ Access to local databases on any server ■ ID must have sufficient access to the server and to the database to perform the action.	■ Unrestricted access to files, change system time, call .DLLs, APIs, etc.
Server Agent	■ Can only open databases on that server ■ The signature of the user who last modified the script is used during the ACLcheck when the agent runs. ■ Server document controls when an agent can run and which LotusScript features it can use on that server.	■ Server document controls whether an agent can run script that can manipulate system settings and files.

Lotus Language Implementation

Table 1.3 outlines the programmable objects in Notes. The third column specifies whether the object supports scripts, formulas, or both. The fourth column shows you where in the Notes user interface to design and activate the scripts and formulas.

Table 1.3 Programmable objects in Notes

Scope	Notes Object	Type	How to Design and Activate
Workspace	SmartIcons®	Formula	To design, choose File . . . Tools . . . SmartIcons. Click the icon, Edit Icon, and then Formula. Activated when the user clicks the icon
Database	Replication formula	Formula	To design, choose File . . . Database . . . Properties. Click Basics, Replication Settings, and then Advanced. Click Only replicate selected documents and the @ symbol. Activated when replication occurs
	Agent	Script Formula	To design, choose View . . . Agents. Select the agent and choose Actions . . . Edit Agent, or choose Create . . . Agent. Click Simple Action, Script, or Formula for Run. Activated according to a schedule, when mail arrives, when a document is modified, or when the user chooses Actions . . . NameOfAgent, or chooses (while in View . . . Agents) Actions . . . Run Agent
	Event	Script Formula	To design, choose View . . . Design and click Other in the Design pane. Double-click Database Script. Select Postopen, Postdocumentdelete, Queryclose, Querydocumentdelete, or Querydocumentundelete in the Event box. Activated when the event occurs
	Event: Initialize/ Terminate	Script	Initialize or Terminate in the Event box; activated when the event occurs

continues

Table 1.3 Continued

Scope	Notes Object	Type	How to Design and Activate
View or Folder Design	Form formula	Formula	To design, choose View . . . Design and click Views or Folders in the Design pane. Double-click the view or folder, or choose Create . . . View or Folder and double-click the new view or folder. Choose Design . . . View or Folder Properties, click the propeller head, and then click Formula Window. Activated when a document is opened from the view or folder
	Selection formula	Formula	To design, choose View . . . Design and click Views or Folders in the Design pane. Double-click the view or folder, or choose Create . . . View or Folder and double-click the new view or folder. Select View or Folder Selection for Define and click Formula for Run. Activated when the view or folder is opened
	Column formula	Formula	To design, choose View . . . Design and click Views or Folders in the Design pane. Double-click the view or folder, or choose Create . . . View or Folder and double-click the new view or folder. Select the column for Define, and click Formula for Run. Activated when the view or folder is opened

Scope	Notes Object	Type	How to Design and Activate
	Action	Script Formula	To design, choose View . . . Design and click Views or Folders in the Design pane. Double-click the view or folder, or choose Create . . . View or Folder and double-click the new view or folder. Choose View or Folder . . . Action Pane and double-click the action, or choose Create . . . Action. Click Simple action, Formula, or Script for Run. Activated when the user chooses Actions . . . NameOfAction or clicks the button
	Hide action formula	Formula	To design, choose View . . . Design and click Views or Folders in the Design pane. Double-click the view or folder, or choose Create . . . View or Folder and double-click the new view or folder. Choose View or Folder . . . Action Pane and double-click the action, or choose Create . . . Action. Click Hide and Hide action if formula is true. Activated when the view or folder is opened
	Event	Script Formula	To design, choose View . . . Design and click Views or Folders in the Design pane. Double-click the view or folder, or choose Create . . . View or Folder and double-click the new view or folder. Select the view or folder in the Define box. Select Queryopen, Postopen, Regiondoubleclick, Queryopendocument, Queryrecalc, Queryaddtofolder, Querypaste, Postpaste, Querydragdrop, PostDragDrop, Queryclose in the Event box. Activated when the event occurs
	Event	Script	Initialize or Terminate in the Event box. Activated when the event occurs

continues

9

Table 1.3 Continued

Scope	Notes Object	Type	How to Design and Activate
Form Design	Window title formula	Formula	To design, choose View . . . Design and click Forms in the Design pane. Double-click the form, or choose Create . . . Design . . . Form. Select the name of the form in the Define box. Select Window Title in the Event box. Activated when a document based on the form is opened
	Section title formula	Formula	To design, choose View . . . Design and click Forms (or Subforms) in the Design pane. Double-click the form (or subform), or choose Create . . . Design . . . Form (or Subform). Select a standard section, or choose Create . . . Section . . . Standard and select the new standard section. Choose Section . . . Section . . . Properties and click Formula. Activated when a document based on the form is opened
	Section access formula	Formula	To design, choose View . . . Design and click Forms (or Subforms) in the Design pane. Double-click the form (or subform), or choose Create . . . Design . . . Form (or Subform). Select a controlled access section, or Choose Create . . . Section . . . Controlled Access. Click the Formula tab. Activated when the section is accessed
	Insert subform formula	Formula	To design, choose View . . . Design and click Forms in the Design pane. Double-click the form, or choose Create . . . Design . . . Form. Choose Create . . . Insert Subform. Click Insert subform based on formula. Activated when a document based on the form is opened

Scope	Notes Object	Type	How to Design and Activate
	Hide paragraph formula	Formula	To design, choose View . . . Design and click Forms (or Subforms) in the Design pane. Double-click the form (or subform), or choose Create . . . Design . . . Form (or Subform). Choose Text . . . Text Properties, click the window blind, and select Hide Paragraph if formula is true. **Activated when the text is accessed**
	Action	Script Formula	To design, choose View . . . Design and click Forms (or Subforms) in the Design pane. Double-click the form (or subform), or choose Create . . . Design . . . Form (or Subform). Choose View . . . Action Pane and select the action, or choose Create . . . Action. Click Simple action, Formula, or Script. **Activated when the user chooses Actions . . . NameOfAction or clicks the button**
	Hide action formula	Formula	To design, choose View . . . Design and click Forms (or Subforms) in the Design pane. Double-click the form (or subform), or choose Create . . . Design . . . Form (or Subform). Choose View . . . Action Pane and select the action, or choose Create . . . Action. Click the Hide tab. Select Hide action if formula is true. **Activated when a document based on the form is opened**

continues

Table 1.3 Continued

Scope	Notes Object	Type	How to Design and Activate
	Event	Script Formula	To design, choose View . . . Design and click Forms (or Subforms) in the Design pane. Double-click the form (or subform), or choose Create . . . Design . . . Form (or Subform). Select the form in the Define box. Select Queryopen, Postopen, Postrecalc, Querysave, Querymodechange, Postmodechange, Queryclose in the Event box. Activated when the event occurs
	Event	Script	Initialize or Terminate in the Event box. Activated when the event occurs
	Button	Script Formula	To design, choose View . . . Design and click Forms (or Subforms) in the Design pane. Double-click the form (or subform), or choose Create . . . Design . . . Form (or Subform). Select a button or choose Create . . . Hotspot . . . Button. Click Simple action, Script, or Formula for Run. Activated when the user opens a document based on the form and clicks the button
	Hotspot	Script Formula	To design, choose View . . . Design and click Forms (or Subforms) in the Design pane. Double-click the form (or subform), or choose Create . . . Design . . . Form (or Subform). Select a hotspot, or select some text and choose Create . . . Hotspot . . . Formula Popup or Action Hotspot. Click Script (Action Hotspot only) or Formula for Run. Activated when the user clicks the hotspot text

Scope	Notes Object	Type	How to Design and Activate
Navigator Design	Hotspot	Script Formula	To design, choose View . . . Design and click Navigators in the Design pane. Double-click the navigator, or choose Create . . . Design . . . Navigator. Select an element or choose Create . . . HotspotElement, where the elements are Hotspot Rectangle, Hotspot Polygon, Graphic Button, Button, Textbox, Rectangle, Rounded Rectangle, Ellipse, Polygon, or Polyline. Click Simple action, Script, or Formula for Run. Activated when the user clicks the hotspot
Layout Region Design	Hotspot Formula	Script	To design, go into form or subform design. Select a layout region or choose Create . . . Layout Region . . . New Layout Region. Select a hotspot or choose Create . . . Hotspot . . . Button. Activated when the user clicks the hotspot
Field Design	Default Value formula for editable field	Formula	To design, go into form, subform, or shared field design. In form design, select the field or choose Create . . . Field. Select Default Value in the Event box. Activated when the document is created
	Input Translation formula for editable field	Formula	To design, go into form, subform, or shared field design. In form design, select the field or choose Create . . . Field. Select Input Translation in the Event box. Activated when the document is saved or recalculated
	Input Validation formula for editable field	Formula	To design, go into form, subform, or shared field design. In form design, select the field or choose Create . . . Field. Select Input Validation in the Event box. Activated after input translation

continues

Table 1.3 Continued

Scope	Notes Object	Type	How to Design and Activate
	Value formula for computed field	Formula	To design, go into form, subform, or shared field design. In form design, select the field or choose Create . . . Field. Select Value in the Event box. Activated when the document is saved or recalculated
	Keyword field formula	Formula	To design, go into form, subform, or shared field design. In form design, select the field or choose Create . . . Field. In the properties box, select Use formula for choices in the Choices box. Activated when the field is edited
	Event	Script	To design, go into form, subform, or shared field design. In form design, select the field or choose Create . . . Field. Select Entering, Exiting, Initialize, or Terminate in the Event box. Activated when the event occurs
Rich text field	Button	Script Formula	To design, enter a rich text field in Edit mode. Select the button and choose Button . . . Edit Button, or choose Create . . . Hotspot . . . Button. Click Simple action, Formula, or Script for Run. If you click Script, select Click in the Event box. Activated when the user clicks the button

Scope	Notes Object	Type	How to Design and Activate
	Hotspot	Script Formula	To design, enter a rich text field in Edit mode. Select the hotspot and choose Hotspot Edit Hotspot or choose Create Hotspot Formula Popup or Action Hotspot. Click Formula or Script (Action Hotspot only) for Run. If you click Script, select Click in the Event box. Activated when the user clicks the hotspot text
	Hotspot	Script	Initialize or Terminate in the Even box. Activated when the user clicks the hotspot text
	Section title formula	Formula	To design, enter a rich text field in Edit mode. Select a standard section, or choose Create . . . Section and select the new standard section. Choose Section . . . Section Properties and click Formula. Activated when a document based on the form is opened
	Hide paragraph	Formula	To design, enter a rich text field in Edit mode. Choose Text . . . Text Properties, click the window blind, and select `Hide Paragraph if formula is true`. Activated when text is accessed

LotusScript Events and Execution

Though it is a procedural language, a script performs in response to the occurrence of an event in an object. The scriptable Notes objects are databases, agents, actions, views (folders), documents (forms), fields, buttons, and hotspots. As such, LotusScript is considered to be an event-driven programming environment. The event initiates execution of a predefined subprogram, for example:

- When a button is clicked by a user, the script in the Click event for the button executes.

- If an agent is scheduled to run every day, the script in the Initialize event executes at the scheduled time.

You can also write formulas to respond to database, view (folder), and document events, except for Initialize and Terminate.

You must understand when events occur. By simply understanding which event executes at which time, you can create more robust and sophisticated applications.

As you read through the list of events and the order of their execution, you will note that there are parameters for some events. The parameters are defined as follows:

- *Source* is a reference to the Button (actions, buttons, and hotspots), Field, Document, Navigator, NotesUIDatabase, NotesUIDocument, or NotesUIView object containing the event.

- *Mode* is *1* if the document is opened in Edit mode, 0 if it is in Read mode.

- *Isnewdoc* is true if the document is new.

- *Continue* is a return parameter. Return false to stop the event.

- *Data* is a handle to the OLE2 server's Dispatch/Automation interface.

- *Target* is the name of the folder to which the document is being added.

> **NOTE**
> In some events, such as the form PostOpen event, not only can you use script—but if you click the formula radio button, you can use the formula language instead. When coding events, if you choose to use LotusScript, you cannot use the formula language in the same event. When you click the radio button to switch from script to formula, you will be prompted that your script or formula will be erased.

Database Events

On Database Open

Table 1.4 describes the events that execute when opening a database through the user interface (i.e., when double-clicking a database icon).

Table 1.4

Seq	Object Type	Event Subprogram	Execution
1	View	Initialize	View when it is being loaded during database open
2	View	Queryopen (*source, continue*)	View before it is opened during database open; *source* is not initialized at this point.
3	View	Postopen (*source*)	View after it is opened during database open
4	Database	Initialize	Database when it is being loaded
5	Database	Postopen (*source*)	Database after it is opened

Intermediary Database Events

These events, shown in Table 1.5, can occur after the database Postopen event has occurred. None of these events will occur at the same time.

Table 1.5

Seq	Object Type	Event Subprogram	Execution
N/A	Database	Queryopendocument (*source, continue*)	Database before document is loaded
N/A	Database	Querydocumentdelete (*source, continue*)	Database before a document is deleted
N/A	Database	Querydocumentundelete (*source, continue*)	Database before a document is undeleted

On Database Close

These events, shown in Table 1.6, occur when closing a database through the user interface:

Table 1.6

Seq	Object Type	Event Subprogram	Execution
1	Database	Queryclose (*source, continue*)	Database when it is being closed
2	Database	Terminate	Database when it is being closed

View Events

On View Open

These events, shown in Tables 1.7, 1.8, 1.9, and 1.10, occur when opening a view through the user interface. For example: @Command([OpenView]), @Command([ViewChange]), or opening a view through the view menu will trigger these events.

Table 1.7

Seq	Object Type	Event Subprogram	Objects
1	View	Initialize	View when new one is being loaded
2	View	Queryopen (*source, continue*)	View before new one is opened; *source* is not initialized at this point.
3	View	Postopen (*source*)	View after new one is opened

Intermediary View Events

Table 1.8

Seq	Object Type	Event Subprogram	Objects
N/A	View	Regiondoubleclick (*source*)	View (calendar) when a region is double-clicked
N/A	View	Queryrecalc (*source, continue*)	View before it is refreshed
N/A	View	Queryaddtofolder (*source, target, continue*)	View before a document is added to a folder

Querypaste Event

Table 1.9

Seq	Object Type	Event Subprogram	Objects
1	View	Querypaste (*source, continue*)	View before a document is pasted
2	View	Postpaste (*source*)	View after a paste operation

Querydragdrop Event

Table 1.10

Seq	Object Type	Event Subprogram	Objects
1	View	Querydragdrop (*source, continue*)	View (calendar) before a dragdrop operation
2	View	Postdragdrop (*source*)	View (calendar) after a dragdrop operation

On View Close

These events, shown in Table 1.11, occur when closing a view through the user interface:

Table 1.11

Seq	Object Type	Event Subprogram	Objects
1	View	Queryclose (*source, continue*)	View when it is being closed
2	View	Terminate	View when it is being closed

Form Events

Opening a Document in Read Mode or Creating a Document

These events, shown in Table 1.12, occur when a document is opened or created through the user interface:

Table 1.12

Seq	Object Type	Event Subprogram	Objects
1	Document	Initialize	Document when it is being loaded
2	Document	Queryopen (*source, mode, isnewdoc, continue*)	Document before it is opened; *source* is not fully initialized at this point.
3	Field	Initialize	Field when document is being loaded
4	Document	Postopen (*source*)	Document after it is opened
5	Action, Button, Hotspot	Initialize	Action, button, or hotspot when it is being loaded. These items load on order of appearance in the form.

Opening a Document in Edit Mode

These events, shown in Table 1.13, occur when a document is being opened in Edit mode. For example, if a document is opened from a view in Edit mode, these events are triggered.

Table 1.13

Seq	Object Type	Event Subprogram	Objects
1	Document	Initialize	Document when it is being loaded
2	Document	Queryopen (*source, mode, isnewdoc, continue*)	Document before it is opened; *source* is not fully initialized at this point.
3	Field	Initialize	Field when document is being loaded
4	Field	Entering	First field when form is opened in Edit mode
5	Document	Postopen (*source*)	Document after it is opened
6	Action, Button, Hotspot	Initialize	Action, button, or hotspot when it is being loaded. These load on order of appearance in the form.

Intermediary Form Events

These events, shown in Table 1.14, can occur after the document Postopen event has occurred. None of these events will occur at the same time.

Table 1.14

Seq	Object Type	Event Subprogram	Objects
N/A	Action, Button, Hotspot	ObjectExecute (*source, data*)	Action, button, or hotspot when it is activated by an OLE2 server that is FX/Notesflow-enabled
N/A	Action, Button, Hotspot	Click (*source*)	Action, button, or hotspot when it is selected

Form Recalculation

These events, shown in Table 1.15, are triggered when a form is refreshed (F9). The Postrecalc event can be useful for form-level validation.

Table 1.15

Seq	Object Type	Event Subprogram	Objects
1	Field	Field Input Translation or Computed formula	
2	Field	Field Validation formulas	
3	Document	Postrecalc (source)	Document after it is refreshed

Document Save

These events, shown in Table 1.16, are triggered when a form is about to be saved. For example, @Command([FileSave]) would trigger these events. This series of events is followed by the document saving to disk (if it passes the Field Validation formulas). Remember that the Querysave event occurs *before* the document is saved, not after.

Table 1.16

Seq	Object Type	Event Subprogram	Objects
1	Document	Querysave (*source, continue*)	Document before it is saved
2	Field	Field Input Translation or Computed formula	
3	Field	Field Validation formulas	

Switching Modes

These events, shown in Table 1.17, occur when a document is switched between Read and Edit modes. If a document is opened in Edit mode and is being switched to Read mode, and if any changes have been made to the document, the Document Save events will occur.

Table 1.17

Seq	Object Type	Event Subprogram	Objects
1	Document	Querymodechange (*source, continue*)	Document before changing to or from Edit mode
2	Document	Postmodechange (*source*)	Document after changing to or from Edit mode

Form Close

These events, shown in Tables 1.18, 1.19, and 1.20, will occur on the closing of a document. If a document is opened in Edit mode and changes have been made, the Document Save events will precede these events.

Table 1.18

Seq	Object Type	Event Subprogram	Objects
1	Document	Queryclose (*source, continue*)	Document before it is closed
2	Action, Button, Hotspot	Terminate	Action, button, or hotspot when document is being closed
3	Field	Terminate	Field when document is being closed
4	Document	Terminate	Document when it is being closed

Field Events

Table 1.19

Seq	Object Type	Event Subprogram	Objects
N/A	Field	Initialize	Field when document is being loaded
N/A	Field	Entering	Field when entered in Edit mode
N/A	Field	Exiting (*source*)	Field when it is exited in Edit mode
N/A	Field	Terminate	Field when document is being closed

Agent Events

Table 1.20

Seq	Object Type	Event Subprogram	Objects
1	Agent	Initialize	Agent when document is being loaded
2	Agent	Terminate	Agent when document is being closed

Programming with LotusScript

The Programming Pane

All scripts are entered into the Programming Pane by choosing the Run: Script radio button, as shown in Figure 1.2.

Figure 1.2 Choose the Run: Script radio button

The Define box reflects the current focus in the window, such as a field or button, and so does not typically need changing. In some windows, for example in the Agent Builder window, the Define box is absent because you do not have a choice of objects, as shown in Figure 1.3.

The Define box also includes the item (Globals), which gives you access to a script where you can define subprograms and declare variables that are available to all objects in the current document.

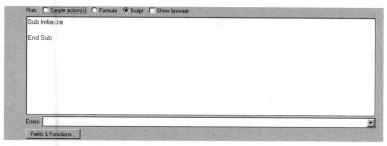

Figure 1.3   Agent Builder window

Depending on the object, the following events are available from the Event box:

- (Options) applies to all scriptable objects. This event provides an area for statements such as Use, Option, Const, and Def.

- (Declarations) applies to all scriptable objects. This event provides an area where you view and declare variables and types that are available to all events in the object. Use the (Declarations) event for the (Globals) object to view and declare variables and types for all events—for all objects in the document.

- Initialize and Terminate apply to all scriptable objects. Loading an object activates its Initialize event. Closing an object activates its Terminate script. Except for agents, these events should only be used for housekeeping tasks, such as loading DLLs. In agents, these are the only available events—and typically, Initialize is used for the code that executes when the user runs the agent.

- The events that start with Query and Post occur in that sequence. The Query event occurs just before an operation, and you can cancel the operation by setting the *continue* parameter to False. The Post event occurs just after the operation.

- The remaining events occur as the result of an operation, such as entering a field.

Programming Pane Browser

To show the Programming Pane Browser, click the Show Browser check box. The Browser will appear to the right of the Programming Pane, as shown in Figure 1.4.

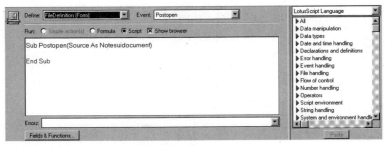

Figure 1.4   Programming Pane Browser

The browser provides the syntax for the following categories of components: the LotusScript language, Notes classes, Notes constants, Notes subroutines and functions, Notes variables, and OLE classes. The default category for the Browser when the Script radio button is selected is the LotusScript Language category. To expand the component categories, click the twisties. To put an entry into the script area, select the entry in the browser and click Paste.

Script Error Message Box

If errors in your LotusScript exist, they appear in the message box at the bottom of the Programming Pane in the order in which they occurred, as shown in Figure 1.5. If no errors exist, the message box is empty. LotusScript messages have the format *object*: *event*: *line #*: *error message*. When you select an error from the box, the corresponding line in the script is highlighted.

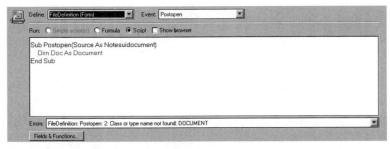

Figure 1.5   Errors message box

Compiling LotusScript

All scripts are compiled when you do the following steps:

- Click off the Programming Pane. This step is a partial compilation. Syntax and other per-line errors are reported at this time.

- Save the design of the container object. For example, when you save a form, all of the script written in the form will compile at that time, assuming there are no errors. This step is a complete compilation. All remaining compile-time errors are reported at this time.

The amount of code that gets compiled depends on your activity in the container object. For example, if you are working within a form and you change code in the (Globals) definition, all scripts associated with the (Globals) code, all code associated with the form (script libraries), and all objects on the form are recompiled. Scripts in shared fields and subforms are not recompiled, because they are not on the form. If you change the code for an object on a form, all events associated with the object are recompiled, but nothing else happens.

If your script contains errors, you will be prompted to correct them when you try to save the container object. The message box shown in Figure 1.6 will appear.

Figure 1.6   Script error message box

If you choose Yes, your cursor will be brought to the last compile error in the last object compiled in your script. If you choose no, the object you are currently scripting will not save.

Saving LotusScript

Because any script you write is part of the overall container object design, such as a form for example, when you save the container object you also compile and save the script.

Debugging LotusScript

The interactive debugger in Notes is an invaluable tool in writing script. The debugger has features that enable you to step through your code, set breakpoints and see the values that your variables and objects contain.

Two types of script errors exist. First, there are compile errors. Compile errors occur due to naming or syntax problems. These errors occur in the Errors combo box, when you move past a line or try to compile the script. Second, there are run-time errors. These errors are unpredictable at compile-time. They include errors such as not referring to a proper variable name when referring to an object, not declaring variables before assigning values to them, data type-mismatches or divide by zero errors.

These types of errors can often be found using the debugger. There is no way, however, for the debugger to find errors in programming logic.

Debug Mode

To enable the debugger, choose File . . . Tools . . . Debug LotusScript from the menu bar. The debugger runs in the background, waiting for an executable script to run. When the event occurs, the debugger comes to the foreground to take control of the script execution. The debugger utilities appear as three tabbed panels in the bottom pane of the debugger. The three tabbed panels are the Breakpoints panel, the Variables panel, and the Output panel. Figure 1.7 shows the debugger dialog box with the Variables panel showing.

Using the Breakpoints Panel

Click the Breakpoints tab or choose Debug . . . Breakpoints from the menu bar to access the breakpoints panel (see Figure 1.7). The Breakpoints panel displays the current breakpoints in the format *object*: *event*: *line*. If the breakpoint is disabled, (Disabled) is appended to the display.

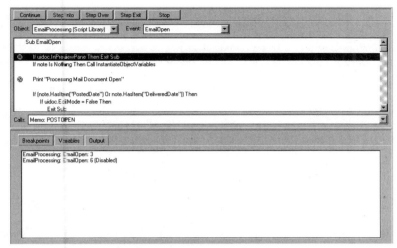

Figure 1.7 Debugger dialog box with Variables panel

Using the Variables Panel

Click the Variables tab or choose Debug . . . Variables from the menu bar to access the variables panel (see Figure 1.8). The variables defined for the procedure appear in a three-column display, showing the name, value, and data type of each variable. To view array or type members, click the twistie to the left of the variable name.

To change the value of a variable, do the following steps, as shown in Figure 1.9:

1. Select the variable

2. Enter the value in the New Value box

3. Click the green check mark

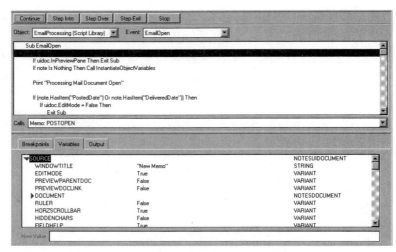

Figure 1.8 Breakpoints panel

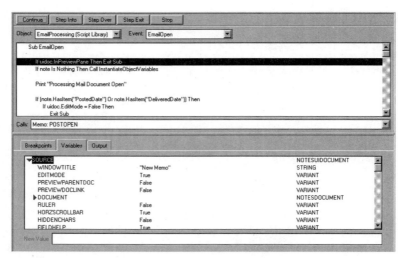

Figure 1.9 Changing the value of a variable

Using the Output Panel

Click the Output tab or choose Debug . . . Output to access the output panel, as shown in Figure 1.10. Script output, for example, from the Print statement, goes to the output panel. You can

- View the output
- Clear the output panel by clicking Clear All
- Copy selected output by choosing Edit . . . Copy. Choose Edit . . . Select All to select all the text in the output panel

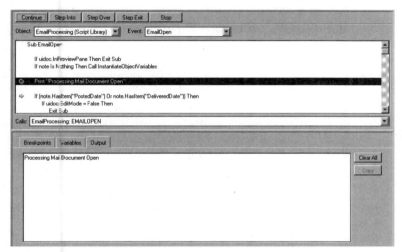

Figure 1.10 Output panel

Selecting a Subprogram

As you execute the script, the current subprogram appears in the debugger window.

The Calls box contains a list of the subprograms currently in the execution stack in order of execution, with the currently executing subprogram at the top of the list. Subprograms are listed as *object*: *event*. For example: `Calculate_totals: CLICK`.

If you select a subprogram from the list, its script appears in the debugger window. If you select the subprogram that is currently executing, the current pointer points to the statement about to be executed. If you select another subprogram, the current pointer points to the statement that calls the next subprogram in the stack.

Stepping through a Script

The debugger provides the following facilities for stepping through a script:

- *Step into a subprogram.* To execute the current statement and step to the next statement, stepping into the subprogram if the current statement calls a subprogram, click the Step Into action bar button or choose Debug . . . Step Into. Step Into proceeds to the next statement in the program. If the current statement calls a subprogram, the debugger displays the code for the subprogram and sets the current line to the first executable statement in the subprogram. If no source code is available for the subprogram (because it is an external file), Step Into behaves similarly to Step Over.

- *Step over a subprogram.* To execute the current statement and step to the next statement, stepping over the subprogram if the current statement calls a subprogram, click the Step Over action bar button or choose Debug . . . Step Over. Step Over proceeds to the next statement in the current program unit. If the statement calls a subprogram, the debugger executes the entire subprogram as if it were a single statement—and sets the current line to the next statement in the calling program unit.

- *Exit from a subprogram.* To execute the remaining statements in a subprogram and step to the next statement in the calling program unit, click the Step Exit action bar button or choose Debug . . . Step Exit. Step Exit continues executing the current subprogram and stops in the subprogram that called it at the line following the call. If the subprogram was not called by another subprogram, execution continues to the next breakpoint or to completion.

Debugging with Breakpoints

A breakpoint interrupts script execution just before the statement at which the breakpoint is set. While script execution is interrupted, you can examine and modify the values of variables and use other debugger commands.

After you set a breakpoint, you can permanently clear it, temporarily disable it, or enable it again. Breakpoints are displayed as red stop signs when enabled—and red stop signs with yellow slashes when disabled.

For a statement continued over multiple lines, the last line is highlighted during stepping and stopping on a breakpoint. To set, disable, enable, or clear a multi-line statement, you must select the last line.

 NOTE
A breakpoint remains in a script until the application session ends or you explicitly clear the breakpoint.

The debugger provides the following breakpoint facilities:

- *Set a breakpoint.* Select a statement at which no breakpoint is currently set. Double-click or choose Debug . . . Set/Clear Breakpoint.

- *Clear a breakpoint.* Select a statement at which a breakpoint is currently set. Double-click once if the stop sign has a yellow slash or twice if the stop sign is solid red, or choose Debug . . . Set/Clear Breakpoint. To clear all breakpoints from all scripts in the active document, choose Debug . . . Clear All Breakpoints.

- *Disable a breakpoint.* Select a statement at which an enabled breakpoint (solid red stop sign) is set. Double-click or choose Debug . . . Disable Breakpoint. To disable all breakpoints from all scripts in the active document, choose Debug . . . Disable All Breakpoints.

- *Enable a breakpoint.* Select a statement at which a disabled (red stop sign with yellow slash) breakpoint is set. Double-click twice or choose Debug . . . Enable Breakpoint. To enable all breakpoints from all scripts in the active document, choose Debug . . . Enable All Breakpoints.

- *Continue script execution.* To start executing the current script or to resume execution after the script is interrupted at a breakpoint, click the Continue action bar button or choose Debug . . . Continue.

Stopping Script Execution

To stop script execution while the debugger is open, choose Debug . . . Stop. All scripts that are at a breakpoint are stopped as if the end of the scripts were reached, and the debugger closes.

Sample Questions

1. *Objective:* Where to use scripts and formulas

   When coding in a Notes object

   a. Script is available for use in all objects.
   b. Script and the @Formula language is available for use in all objects.
   c. Whether script or the formula language is available for coding depends on the object.
   d. Script is available for use in only in forms and in fields.

2. *Objective:* When to use scripts and formulas

   In general

   a. Script is best used for working within the object that the user is currently processing, for example, to return a default value to a field or determine selection criteria for a view.
   b. Script is best used in validation formulas.
   c. Script is best used for accessing existing objects—for example, to change a value in one document based on values in other documents.
   d. Script is best used when a Simple Action cannot complete the task.

3. *Objective:* Advantages of using script

   One advantage of LotusScript is

   a. It transparently handles multi-value fields.
   b. It has robust text manipulation methods.
   c. Complicated control structures must be written to manipulate multi-value fields.
   d. It enables access to many database and document properties.

4. *Objective:* LotusScript implementation

   Script cannot be used in

   a. Fields
   b. Agents
   c. Action
   d. Keyword list formulas

5. *Objective:* LotusScript execution

   The form Query save event occurs

   a. After the document has been saved to disk
   b. Before the document is saved and before the field translation and input validation formulas
   c. Before field validation but after input validation
   d. After the input translation formulas

6. *Objective:* Compiling LotusScript

   A partial LotusScript compilation occurs when

   a. Clicking off the programming pane from a scriptable object
   b. Saving the container object
   c. After each line of script is written
   d. A database that contains LotusScript is closed.

7. *Objective:* Compiling LotusScript

   If you change the code for an object on a form and save the container object

   a. All the code in the container object is recompiled.
   b. Only the changed code is recompiled.
   c. Nothing is recompiled.
   d. All events associated with the object are recompiled but nothing else happens.

8. *Objective:* Debugging LotusScript

   The debugger has features that enable you to

   a. Step through your code
   b. Step through your code, set breakpoints, and see the values that your variables and objects contain
   c. Set breakpoints and see the values that your variables and objects contain
   d. Step through your code, set breakpoints, see the values that your variables and objects contain and edit your code while in the debugger

9. *Objective:* Debugging LotusScript

   The two types of LotusScript errors that the debugger will help you debug are

   a. Divide by zero errors and syntax
   b. Run-time errors and logic errors
   c. Compile errors and type-mismatch errors
   d. Compile errors and run-time errors

10. *Objective:* Using the Debugger

    The debugger provides the following breakpoint facilities

    a. Setting and clearing breakpoints
    b. Setting, clearing and enabling breakpoints
    c. Enabling and disabling breakpoints
    d. Setting, clearing, enabling, disabling breakpoints and continuing execution of the script

Sample Answers

1. *Answer:* c

   Where you can use script depends on the object. Not all codable objects in Notes enable the use of LotusScript. For example, SmartIcons do not enable the usage of script.

2. *Answer:* c

   Script is best used for accessing existing objects—for example, to change a value in one document based on values in other documents. Script is also best used when looping functionality is needed to complete a task.

3. *Answer:* d

   Script, through its object classes, enables the retrieval and setting of many database and document properties, such as the database title or the document IsResponse property.

4. *Answer:* d

   Only Simple Actions and the @Formula language can be used in keyword list formulas.

5. *Answer:* b

   The Querysave event occurs before the field translation and input validation formulas.

6. *Answer:* a

   A partial LotusScript compilation occurs when you click off the programming pane from a scriptable object. A full LotusScript compilation occurs when the container object is saved.

7. *Answer:* d

   If you change the code for an object on a form, all the events associated with the object are recompiled, but nothing else happens. For example, if you have written a subroutine and change the code, any event, subroutine, or function that calls it will be recompiled.

8. *Answer:* b

   In the LotusScript debugger, you can step through your code, set breakpoints, and see the values that your variables and objects contain. You cannot edit your code while you are in the debugger.

9. *Answer:* b

   The LotusScript debugger will help you debug compile and run-time errors. Divide by zero and type-mismatch errors are types of compile errors. The debugger will not help you with coding logic errors.

10. *Answer:* b

   In the LotusScript debugger, you can set, clear, enable, and disable breakpoints and continue the execution of the script by clicking the continue button.

CHAPTER 2

LotusScript Language

As with any programming language, you must understand the basic elements of the language before you can begin programming effectively. This chapter begins by introducing you to the available data types used in LotusScript. You will notice that you have more data type choices in LotusScript than you do when physically creating a field on a form in Lotus Notes. This chapter then introduces you to the concepts of variable declarations, construction rules, statements and functions, and finally arrays. This chapter is an important chapter to understand, because it is the foundation for all the script you will write.

TIP
Understanding this chapter is critical to passing the exam. If you do not know the basics, you will get tripped up when having to choose between two similar answers. You should pay particular attention to the section of this chapter that covers arrays. This should not overshadow the importance of the rest of the chapter. There are questions that relate to all sections of this chapter.

Chapter Objectives

The objectives of this chapter are to increase your understanding of

- Looping (Iterating)
- Logic and program flow
- Options (alternation/branching structures)
- Functions and subroutines
- Array processing
- Option base statements
- Declarations
- Variable initialization
- Error-handling

LotusScript Data Types

LotusScript recognizes the following scalar (numeric and string) data types, as shown in Table 2.1.

Table 2.1   Recognized Scalar Data Types

Data Type	Suffix	Value Range	Size
Integer	%	−32,768 to 32,767; Initial value: 0	2 bytes
Long	&	−2,147,483,648 to 2,147,483,647 Initial value: 0	4 bytes
Single	!	−3.402823E+38 to 3.402823E+38 Initial value: 0	4 bytes
Double	#	−1.7976931348623158+308 to 1.7976931348623158+308 Initial value: 0	8 bytes

Data Type	Suffix	Value Range	Size
Currency	@	–922,337,203,685,477.5807 to 922,337,203,685,477.5807 Initial value: 0	8 bytes
String	$	(String length ranges from 0 to 32K characters) Initial value: " " (empty string)	(2 bytes/ character)

The Suffix column in the table tells you which character you can use to implicitly declare variables. Figure 2.1 shows an example.

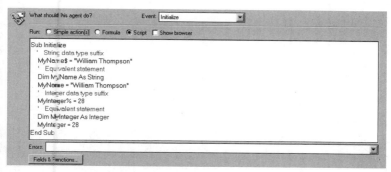

Figure 2.1   Suffixes used to declare variables

Notice that the equivalent statements in the example need two lines to achieve the same result. The equivalent statements explicitly declare the variables.

Besides the scalar data types, LotusScript supports the following additional data types and data structures, as shown in Table 2.2.

Table 2.2 LotusScript Supported Data Types and Data Structures

Data Type or Structure	Description	Size
Array	An aggregate set of elements having the same data type. An array can comprise up to eight dimensions whose subscript bounds can range from -32768 to 32767. Initial value: Each element in a fixed array has an initial value appropriate to its data type.	Up to 64KB
List	A one-dimensional aggregate set whose elements have the same data type and are referred to by name, rather than by subscript	Up to 64KB
Variant	A special data type that can contain any scalar value, array, list, or object reference. Initial value: EMPTY	16 bytes
User-defined data type	An aggregate set of elements of possibly disparate data types. Comparable to a record in Pascal or a struct in C. Initial value: Member variables have initial values appropriate to their data types.	Up to 64KB
Class	An aggregate set of elements of possibly disparate data types together, with procedures that operate on them. Initial value: When you create an instance of a class, LotusScript initializes its member variables to values appropriate to their data types and generates an object reference to it.	
Object reference	A pointer to an OLE Automation object or an instance of a product class or user-defined class. *-Initial value: NOTHING	4 bytes

In each of the preceding tables, the specified storage size is platform-independent.

Variable Declarations

To declare variables in LotusScript, you need to use the Dim statement. You can declare a variable name either explicitly or implicitly. The Dim statement explicitly declares a name. A name is declared implicitly if it is used (referred to) when it has not been explicitly declared, or when it is not declared as a Public name in another module being used by the module to where the name is referred. You can prohibit implicit declarations by including the statement Option Declare in the (Options) area of your script.

Explicit Declaration of Variables

A name is explicitly declared if it is declared by a Dim statement. Variable declarations are usually done in the (Globals)(Declarations). You can Dim variables using the following syntax:

```
{ Dim } variableDeclaration [ , variableDeclaration ] . . .
```

Figure 2.2 shows an example of an explicit (Declaration).

Figure 2.2   Explicit (Declaration)

Implicit Declaration of Variables

A name is implicitly declared if it is used (referred to) when it has not been explicitly declared by a Dim statement—and also has not been declared as a Public name in another module that is used by the module to where the name is referred.

LotusScript declares the name as a scalar variable, establishing its data type by using the following rules:

- If the name is suffixed by a data type suffix character, this determines the variable data type.

- If no data type suffix character is specified in the first use of the name, the data type is determined by the applicable Def*type*, if any.

- If there is no suffix character and no applicable Def*type*, the variable is of type Variant.

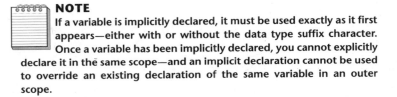 **NOTE**
If a variable is implicitly declared, it must be used exactly as it first appears—either with or without the data type suffix character. Once a variable has been implicitly declared, you cannot explicitly declare it in the same scope—and an implicit declaration cannot be used to override an existing declaration of the same variable in an outer scope.

Figure 2.3 shows an example of an implicit (Declaration).

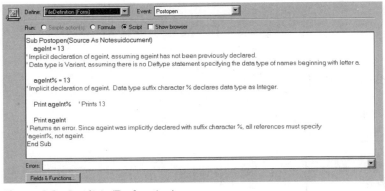

Figure 2.3 Implicit (Declaration)

Initializing Variables

Declaring a variable also initializes it to a default value and follows these rules:

- Scalar variables are initialized according to their data type:

  - Numeric data types (integer, long, single, double, currency) Zero (0)

- Variants: EMPTY
- Fixed-length strings: A string filled with the null character Chr(0)
- Variable-length strings: The empty string (" ")

- Object reference variables are initialized to NOTHING, unless New is specified in the variable declaration.

  - Each member of a user-defined data type variable is initialized according to its own data type.

  - Each element of an array variable is initialized according to the array's data type.

  - A list variable has no elements when it is declared, so there is nothing to initialize.

Visibility of Declarations

The visibility of declarations follows these rules.

- The default visibility for a declaration at the module level is Private, unless Option Public has been specified.

- The default visibility for a variable declaration within a class is Private.

- Public and Private can only be used to declare variables in module or class scope.

- Variables declared within a procedure are automatically Private; however, members of user-defined data types are automatically Public (these cannot be changed).

 NOTE
For further discussion on this topic, see Chapter 3, "Reusing Objects."

Construction Rules

Statement Construction Rules

The following rules govern the construction of statements in a script:

- The statements of a script are composed of lines of text. Each text element is a LotusScript keyword, operator, identifier, literal, or special character.

- The script can include blank lines without effect.

- The text on a line can begin at the left margin or can be indented without affecting the meaning.

- Within a statement, elements are separated with white space (spaces or tabs). Where white space is legal, extra white space can be used to make a statement more readable, but it has no effect.

- Statements typically appear one to a line. A new line marks the end of a statement, except for a block statement. The beginning of the next line starts a new statement.

- Multiple statements on a line must be separated by a colon (:). If you use this feature, Notes will automatically separate the statements onto separate lines.

- A statement, except for a block statement, must appear on a single line—unless it includes the line-continuation character underscore (_), preceded by white space.

- The line continuation character must appear at the end of a line to be continued, preceded by at least one space or tab. Only white space or in-line comments (those preceded with an apostrophe) can follow the underscore on the line. Line continuation within a literal string or a comment is not permitted.

An example is shown in Figure 2.4.

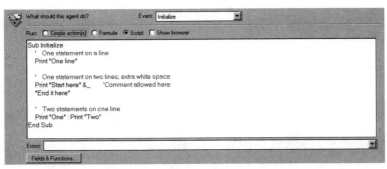

Figure 2.4   Line continuation example

Literal String Construction Rules

The following rules govern the construction of literal strings in a script.

- A literal string in LotusScript is a string of characters enclosed in one of the following sets of delimiters:
  - A pair of double quotation marks (" ")
  - A pair of vertical bars (| |)
  - Open and closed braces ({ })
- A literal string can include any character.
- Strings enclosed in vertical bars or braces can span multiple lines.
- To include one of the closing delimiter characters ", | , or } as text within a string delimited by that character, double it.
- The empty string has no characters at all; rather, it is represented by " ".
- Strings delimited by vertical bars, braces, or double quotation marks cannot be nested.

    An example is shown in Figure 2.5.

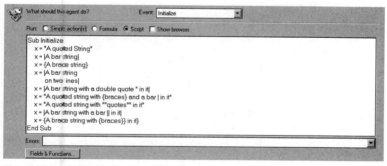

Figure 2.5 Literal string continuation

Literal Number Construction Rules

The following rules, shown in Table 2.3, govern the construction of literal numbers in a script.

Table 2.3 Literal Number Construction Rules

Kind of Literal	Example	Description
Decimal integer	777	The legal range is the range for Long values. If the number falls within the range for Integer values, its data type is Integer; otherwise, its data type is Long.
Decimal	8	The legal range is the range for Double values. The number's data type is Double.
Scientific notation	7.77E+02	The legal range is the range for Double values. The number's data type is Double.
Binary integer	&B1100101	The legal range is the range for Long values. A binary integer is expressible in 32 binary digits of zero or one. Values of &B100000 . . . (31 zeros) and larger represent negative numbers.
Octal integer	&O1411	The legal range is the range for Long values. An octal integer is expressible in 11 octal digits of zero to seven. Values of &O20000000000 and larger represent negative numbers. Values of &O40000000000 and larger are out of range.
Hexadecimal integer	&H309	The legal range is the range for Long values. A hexadecimal integer is expressible in eight hexadecimal digits of zero to nine and A to F. Values of &H80000000 and larger represent negative numbers.

Identifier Construction Rules

An identifier is the name given to a variable, a constant, a type, a class, a function, a sub, or to a property.

The following rules govern the construction of identifiers in a script:

- The first character in an identifier must be an upper-case or lower-case letter.
- The remaining characters must be letters, digits, or underscore.
- A data type suffix character can be appended to an identifier but is not part of the identifier.
- The maximum length of an identifier is 40 characters, not including the optional suffix character.
- Names are case-insensitive. For example, CurEd is the same as cured.
- Characters with ANSI codes larger than 127 (i.e., characters whose ANSI codes are outside the ASCII range) are legal in identifiers.

Escape Character for Illegal Identifiers

Lotus products and OLE classes might define properties or methods whose identifiers use characters that are not legal in LotusScript identifiers. Variables registered by Lotus products might also use such characters. In these cases, prefix the illegal character with a tilde (~) to make the identifier valid. Figure 2.6 shows an example.

Figure 2.6 Illegal identifiers

Labels in LotusScript

A label gives a name to a statement and is built in the same way as an identifier. A label is followed by a colon (:) and cannot be suffixed with a data type suffix character.

The following statements transfer control to a labeled statement by referring to its label:

- GoSub
- GoTo
- If . . . GoTo
- On Error
- On . . . GoSub
- On . . . GoTo
- Resume

The following rules govern the use of labels in a script, and several examples are shown in Figure 2.7.

- A label can appear at the beginning of a line labeling the first statement on the line.

- A label can appear on a line by itself. This option labels the first statement starting after the line.

- A given statement can have more than one label preceding it, but the labels must appear on different lines.

- A given label cannot be used to label more than one statement in the same procedure.

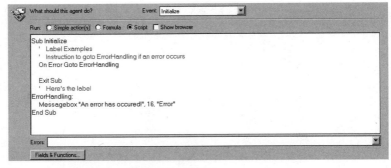

Figure 2.7 Examples of labels in a script

Keywords in LotusScript

A keyword is a word with reserved meaning in the LotusScript language. The keywords name LotusScript statements, built-in functions, built-in constants, and data types. The keywords New and Delete can be used to name subs that you can define in a script. Other keywords are not names but appear in statements—for example, NoCase or Binary. Some of the LotusScript operators are keywords, such as Eqv and Imp.

You cannot redefine keywords to mean something else in a script, with one exception: They can name variables within a type and variables and procedures within a class.

The following list indicates all the LotusScript keywords:

Abs	Access	Acos	ActivateApp
Alias	And	Any	Append
As	Asc	Asin	Atn
Atn2	Base	Beep	Bin
Bin$	Binary	Bind	ByVal
Call	Case	Ccur	CDatCDbl
ChDir	ChDrive	Chr	Chr$
Cint	Class	Clng	Close
Command	Command$	Compare	Const
Cos	Csng	Cstr	CurDir
CurDir$	CurDrive	CurDrive$	Currency
Cvar	DataType	Date	Date$
DateNumber	DateValue	Day	Declare
DefCur	DefDbl	DefInt	DefLng
DefSng	DefStr	DefVar	Delete
Dim	Dir	Dir$	Do
Double	Else	ElseIf	End
Environ	Environ$	EOF	Eqv
Erase	Erl	Err	Error
Error$	Evaluate	Event	Execute
Exit	Exp	FALSE	FileAttr
FileCopy	FileDateTime	FileLen	Fix
For	ForAll	Format	Format$
Fraction	FreeFile	From	Function
Get	GetFileAttr	GoSub	GoTo
Hex	Hex$	Hour	If
IMEStatus	Imp	In	Input
Input$	InputB	InputB$	InputBox

InputBox$	InputBP	InputBP$	InStr
InStrB	InStrBP	Int	Integer
Is	IsArray	IsDate	IsElement
IsEmpty	IsList	IsNull	IsNumeric
IsObject	IsScalar	IsUnknown	Kill
Lbound	Lcase	Lcase$	Left
Left$	LeftB	LeftB$	LeftBP
LeftBP$	Len	LenB	LenBP
Let	Lib	Like	Line
List	ListTag	LMBCS	Loc
Lock	LOF	Log	Long
Loop	Lset	Ltrim	Ltrim$
Me	MessageBox	Mid	Mid$
MidB	MidB$	Minute	MkDir
Mod	Month	Name	New
Next	NoCase	NoPitch	Not
NOTHING	Now	NULL	Oct
Oct$	On	Open	Option
Or	Output	PI	Pitch
Preserve	Print	Private	Property
Public	Put	Random	Randomize
Read	ReDim	Rem	Remove
Reset	Resume	Return	Right
Right$	RightB	RightB$	RightBP
RightBP$	RmDir	Rnd	Round
Rset	Rtrim	Rtrim$	Second
Seek	Select	SendKeys	Set
SetFileAttr	Sgn	Shared	Shell
Sin	Single	Space	Space$
Spc	Sqr	Static	Step
Stop	Str	Str$	StrCompare
String	String$	Sub	Tab
Tan	Then	Time	Time$
TimeNumber	Timer	TimeValue	To
Today	Trim	Trim$	TRUE
Type	TypeName	Ubound	
Ucase	Ucase$	Uchr	Uchr$
Uni	Unicode	Unlock	Until
Use	UseLSX	Ustring	Ustring$
Val	Variant	Weekday	Wend
While	Width	With	Write
Xor	Year	Yield	

Special Characters

LotusScript uses special characters, such as punctuation marks, for several purposes:

- To delimit literal strings
- To designate variables as having particular data types
- To punctuate lists, such as argument lists and subscript lists
- To punctuate statements
- To punctuate lines in a script

NOTE
Special characters within literal strings are treated as ordinary text characters.

Table 2.4 summarizes the special characters used in LotusScript.

Table 2.4 LotusScript Special Characters

Character	Usage
" (quotation mark)	Opening and closing delimiter for a literal string on a single line
| (vertical bar)	Opening and closing delimiter for a multi-line literal string. To include a vertical bar in the string, use double bars (||).
{ } (braces)	Delimits a multi-line literal string. To include an open brace in the string, use a single open brace ({). To include a close brace in the string, use double close braces (}}).
: (colon)	(1) Separates multiple statements on a line. (2) When following an identifier at the beginning of a line, a colon designates the identifier as a label.

continues

Table 2.4 Continued

Character	Usage
$ (dollar sign)	(1) When suffixed to the identifier in a variable declaration or an implicit variable declaration, the dollar sign declares the data type of the variable as String. (2) When prefixed to an identifier, the dollar sign designates the identifier as a product constant.
% (percent sign)	(1) When suffixed to the identifier in a variable declaration or an implicit variable declaration, the percent sign declares the data type of the variable as Integer. (2) When suffixed to either the identifier or the value being assigned in a constant declaration, the percent sign declares the constant's data type as Integer. (3) Designates a compiler directive, such as %Rem or %If
& (ampersand)	(1) When suffixed to the identifier in a variable declaration or an implicit variable declaration, the ampersand declares the data type of the variable as Long. (2) When suffixed to either the identifier or the value being assigned in a constant declaration, the ampersand declares the constant's data type as Long. (3) Prefixes a binary (&B), octal (&O), or hexadecimal (&H) number. (4) Designates the string link operator in an expression
! (exclamation point)	(1) When suffixed to the identifier in a variable declaration or an implicit variable declaration, the exclamation point declares the data type of the variable as Single. (2) When suffixed to either the identifier or the value being assigned in a constant declaration, the exclamation point declares the constant's data type as Single.

Character	Usage
# (pound sign)	(1) When suffixed to the identifier in a variable declaration or an implicit variable declaration, the pound sign declares the data type of the variable as Double. (2) When suffixed to either the identifier or the value being assigned in a constant declaration, the pound sign declares the constant's data type as Double. (3) When prefixed to a literal number or a variable identifier, the pound sign specifies a file number in certain file I/O statements and functions.
@ (at sign)	(1) When suffixed to the identifier in a variable declaration or an implicit variable declaration, the at sign declares the data type of the variable as Currency. (2) When suffixed to either the identifier or the value being assigned in a constant declaration, the at sign declares the constant's data type as Currency.
* (asterisk)	(1) Specifies the string length in a fixed-length string declaration. (2) Designates the multiplication operator in an expression
( ) (parentheses)	(1) Groups an expression, controlling the order of evaluation of items in the expression. (2) Encloses an argument in a sub or function call that should be passed by value. (3) Encloses the argument list in function and sub definitions and in calls to functions and subs. (4) Encloses the array bounds in array declarations and the subscripts in references to array elements. (5) Encloses the list tag in a reference to a list element

continues

Table 2.4   Continued

Character	Usage
. (period)	(1) When suffixed to a type variable or an object reference variable, the period references members of the type or object. (2) As a prefix in a product object reference, the period designates the selected product object. (3) As a prefix in an object reference within a With statement, the period designates the object referred to by the statement. (4) Designates the decimal point in a floating-point literal value
.. (two periods)	Within a reference to a procedure in a derived class that overrides a procedure of the same name in a base class, two periods specify the overridden procedure.
[ ] (brackets)	Brackets delimit names used by certain Lotus products to identify product objects.
, (comma)	(1) Separates arguments in calls to functions and subs and in function and sub definitions. (2) Separates bounds in array declarations and subscripts in references to array elements. (3) Separates expressions in Print and Print # statements. (4) Separates elements in many other statements
; (semicolon)	Separates expressions in Print and Print # statements
' (apostrophe)	Designates the beginning of a comment. The comment continues to the end of the current line.
_ (underscore)	When preceded by at least one space or tab, the underscore continues the current line to the next line.

White space is needed primarily to separate names and keywords—or to make the use of a special character unambiguous. White space is not needed with most non-alphanumeric operators, and it is sometimes incorrect to use it around a special character, such as a data type suffix character appended to a name.

Statements and Functions

LotusScript keywords are reserved words that represent statements and functions. Statements perform an action without returning a value. Functions perform an action and return a value to a declared variable or to a variant. For example, a statement using Print, shown in Figure 2.8, will not return a value but will perform an action.

Figure 2.8 A statement using Print

A function, such as Inputbox, shown in Figure 2.9, will return a value to a declared variable or to a variant.

Figure 2.9 Inputbox function

If you use the debugger, shown in Figure 2.10, you can see the value returned from the function.

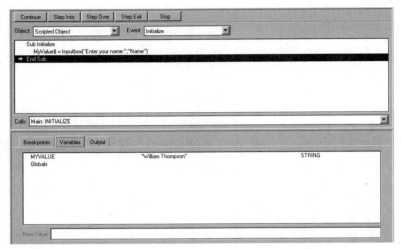

Figure 2.10   Debugger showing the value returned from the function

Control Structures

The structure of a program refers to the sequence in which lines of script execute. If you

- Do not use any branching or looping structures, the lines of script execute in sequence.
- Use branching structures, the program execution jumps to another location in the script.
- Use looping structures, the program execution repeats lines of script until a condition comes true—and then execution continues on the line following the looping structure.

Branching Structures

If . . . Then . . . Else Statement

If statements, shown in Figure 2.11, enable you to test a statement and execute a block of script using this syntax:

```
If condition Then [ statements ] [ Else [ statements ] ]
```

If *condition* is TRUE, the statements following Then, if any, are executed. If *condition* is FALSE, the statements following Else are executed. If no statements follow Then, and there is no Else clause, Then must be followed by a colon (:). Otherwise, LotusScript assumes that the statement is the first line of an If . . . Then . . . Else . . . End If statement. As soon as one of the conditions is met, the statement block is executed, and the control passes to the line following the End If.

An example is shown in Figure 2.11.

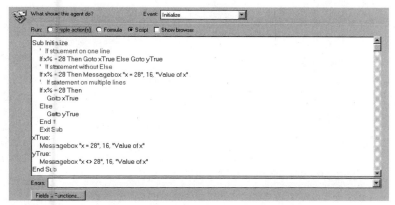

Figure 2.11 If . . . Then . . . Else example

Select Case Statement

Select Case offers a way to structure program flow for many possible responses for a single test expression, choosing the first case that meets the expression using this syntax:

```
Select Case selectExpr
      [ Case condList
            [ statements ] ]
      [ Case condList
            [ statements ] ]…
      [ Case Else
            [ statements ] ]
   End Select
```

You can use Select Case to take user input and select an appropriate function parameter or when interpreting data provided by keywords synonyms.

An example is shown in Figure 2.12.

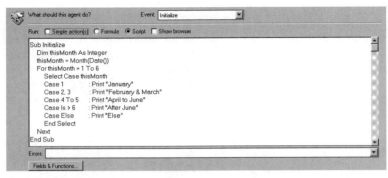

Figure 2.12 Select Case example

On . . . Goto Statement

Similar to a Select Case structure, the On/Goto structure transfers control to a label depending on the value of the expression, using this syntax:

```
On numExpr GoTo label [ , label ] . . .
```

An example is shown in Figure 2.13.

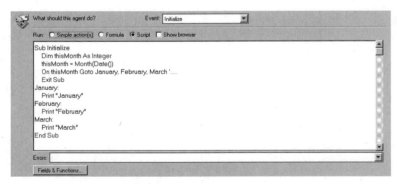

Figure 2.13 On . . . Goto example

In this example, if thisMonth returns a value of one, control of the script will go to the first label listed after the Goto, which is January.

Looping Structures

Do While/Loop and Do Until/Loop Structure

The Do While looping structure loops through a statement block while or until a condition comes true. The syntax is

```
Do [ While | Until condition ]
    [ statements ]
Loop
```

You can exit the loop with an Exit Do statement or a GoTo statement. Exit Do transfers control to the statement that follows the Do . . . Loop block; GoTo transfers control to the statement at the specified label.

If you do not specify a While or Until *condition*, the loop will run forever or until an Exit Do or a GoTo statement is executed within the loop. For example, a Do While loop is shown in Figure 2.14. Figure 2.15 shows a Do Until example.

Figure 2.14   Do While Loop example

Figure 2.15   Do Until example

While/Wend Looping Structure

Executes a block of statements repeatedly while a given condition is true.

Syntax

```
While condition
     [ statements ]
Wend
```

Figure 2.16 shows an example.

Figure 2.16   While/Wend example

ForAll Looping Structure

The ForAll looping structure repeatedly executes a statement block for each element of an array, list, or collection using the syntax:

```
ForAll refVar In ArrayName
     [ statements ]
End ForAll
```

Figure 2.17 shows an example.

When referencing the array, you do not have to specify the location number, because it is implied in the ForAll loop. The previous example creates, loads, and then displays the values of the employee Names. The variable n (a variant, which should not be declared ahead of time) becomes a proxy for the array element and is operated on as though it were the element during that loop through the array.

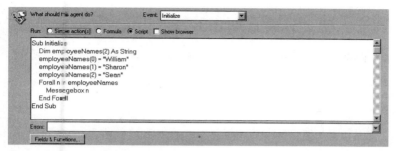

Figure 2.17 ForAll Looping example

For/Next Looping Structure

The For/Next looping structure runs a statement block a specified number of times, using this syntax:

```
For countVar = first To last [ Step increment ]
    [ statements ]
Next [ countVar ]
```

Executing the Loop the First Time

Before the block of statements is executed for the first time, *first* is compared to *last*. If *increment* is positive and *first* is greater than *last*, or if *increment* is negative and *first* is less than *last*, the body of the loop is not executed. Execution continues with the first statement following the For loop's terminator (Next).

Otherwise, *countVar* is set to *first* and the body of the loop is executed.

Executing the Loop More Than Once

After each execution of the loop, *increment* is added to *countVar*. Then *countVar* is compared to *last*. When the value of *countVar* is greater than *last* for a positive *increment*, or less than *last* for a negative *increment*, the loop is complete—and execution continues with the first statement following the For loop's terminator (Next). Otherwise, the loop is executed again.

Exiting the Loop Early

You can exit a For loop early with an Exit For statement or a GoTo statement. When LotusScript encounters an Exit For, execution continues with the first statement following the For loop's terminator (Next). When LotusScript encounters a GoTo statement, execution continues with the statement at the specified label.

Figure 2.18 shows an example.

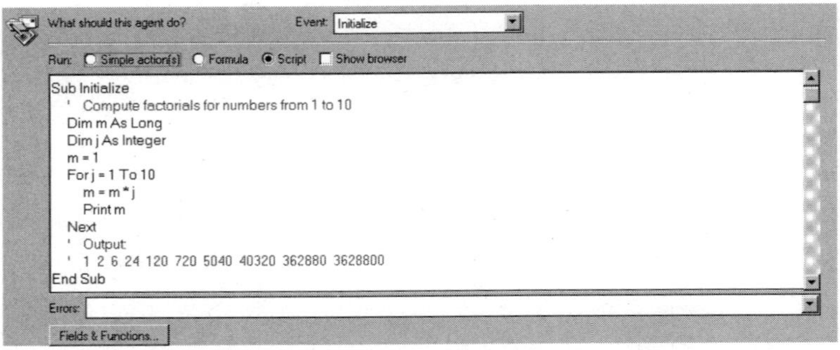

```
Sub Initialize
   '  Compute factorials for numbers from 1 to 10
   Dim m As Long
   Dim j As Integer
   m = 1
   For j = 1 To 10
      m = m * j
      Print m
   Next
   '  Output:
   '  1 2 6 24 120 720 5040 40320 362880 3628800
End Sub
```

Figure 2.18   Example of exiting a loop early

Arrays

An *array* is a named collection of elements of the same data type, where each element can be accessed individually by its position within the collection. If a scalar variable names a single location in memory, an array variable names a series of locations in memory, each holding a value of the same type, integer, or string, for example. The position of an element in an array can be identified by one or more coordinates called *subscripts* (or indexes). The number of subscripts necessary to identify an element is equal to the number of the array's dimensions. In a one-dimensional array, a given element's position can by described by one subscript; however, in a two dimensional array, it takes two subscripts to locate an element. In a three dimensional array, it takes three subscripts—and so on.

An array's elements can be predefined or dynamically reassigned, once you know how many elements the array should contain.

Declaring Arrays

For a fixed array, Dim specifies the type of the array, the number of the array's dimensions, and the subscript bounds for each dimension. Dim allocates storage for the array elements and initializes the array elements to the appropriate value for that data type.

For a dynamic array, Dim only specifies the type of the array. The number of dimensions of the array and the subscript bounds for each dimension are not defined, and no storage is allocated for the array elements. The declaration of a dynamic array must be completed by a ReDim statement.

Static Arrays

The statement in Figure 2.19 creates a static, single-dimension array with two elements: zero and one. The following figure shows what the variable will look like in the debugger.

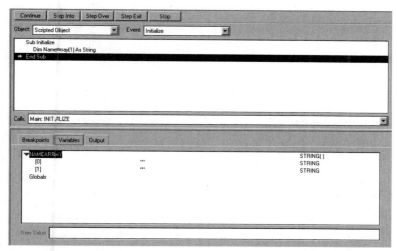

Figure 2.19 Creating a static, single-dimension array with two elements

Arrays can have up to eight dimensions. To declare a four-element array of strings in two dimensions, your declaration would use this syntax:

```
Dim NameArray(1, 1) As String
```

Figure 2.20 shows what the variable will look like in the debugger. Note that the default base (or starting) element for arrays is zero. To change this, you need to use the Option Base statement in the script Options definition. Option Base will enable you to change the default base element. This statement will change the base element to one:

```
Option Base 1
```

Here is what the two dimension array example would look like, with Option Base set to 1, shown in Figure 2.21.

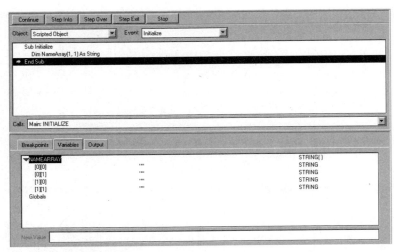

Figure 2.20   The variable in the debugger

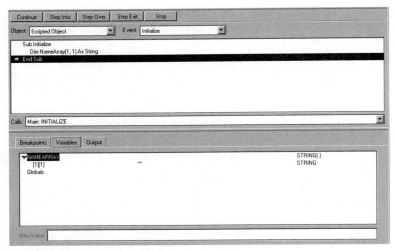

Figure 2.21   Two-dimension array example

Because we have included the Option Base statement in this example, our two-dimensional, four-element array becomes a two-dimensional, one-element array—all because we have changed our default base array element to one.

Arrays can consume 64K of memory, with upper and lower bounds of 32,767 to -32,767. What this means, for example, is that a long, one-dimensional, fixed array (4 bytes for each element) can have up to 16,128 elements.

Dynamic Arrays

To declare a dynamic array, omit the number of elements for the array in the declaration. For example

```
Dim NameArray() As String
```

To change the size of a dynamic array, you must use the ReDim statement to set a new upper bound. The syntax for the ReDim statement is

Syntax

```
ReDim [ Preserve ] arrayName ( bounds ) [ As type][ , arrayName
( bounds ) [ As type ] ]  . . .
```

If you have already declared *arrayName* by using the Preserve parameter, LotusScript preserves the values currently assigned to it. If you omit Preserve, LotusScript initializes all elements of the array, depending on the data type of the array variable.

The following example, shown in Figure 2.22, declares a dynamic array and then resizes it.

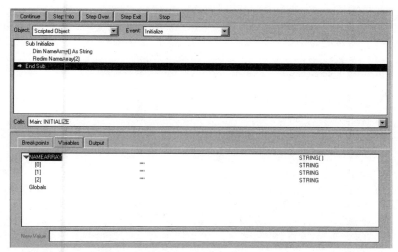

Figure 2.22 Declaring and resizing a dynamic array

Assigning Array Values

To set a value in an array, you must reference the subscript of the desired element. For example, to set both elements of a two element array

```
NameArray(0) = "Bill"
NameArray(1) = "Sharon"
```

Retrieving Array Values

As in assigning array values, when retrieving array values you must reference the desired subscript of the element. For example

```
Messagebox NameArray(0)
```

This command will display the value in the first element in the NameArray variable in a message box.

Determining the Number of Array Elements

Being able to determine the number of array elements is important when you want to be able to add an element to a dynamic array, for example. By using the Lbound and Ubound statements in Lotus-Script, we can determine an array's upper and lower bounds. The following example, shown in Figure 2.23, uses the Ubound statement to add an element to a dynamic array.

Ubound returns the upper-bound subscript for one dimension of an array.

Syntax
```
UBound ( arrayName [ , dimension ] )
```

Elements
`arrayName`   This element is the name of an array.

`dimension`   Optional. An integer argument that specifies the array dimension for which you want to retrieve the upper bound.

Return Value
Ubound returns an Integer.

Lbound returns the lower-bound subscript for one dimension of an array.

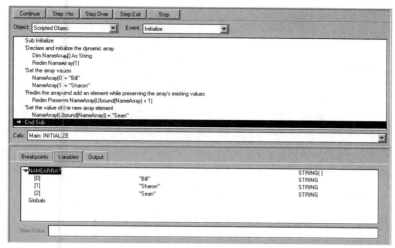

Figure 2.23   Adding an element with a Ubound statement

Syntax

LBound (*arrayName* [, *dimension*])

Elements

arrayName   This element is the name of an array.

dimension   Optional. An integer argument that specifies the array dimension; the default is one.

Return Value

Lbound function returns an Integer.

Recursively Moving through an Array

Typically, you can load and access array elements using two structures:

- For/Next if you now the number of elements ahead of time (or by using the Ubound function if the size of the array is unknown)
- ForAll to access all of the elements

This example, shown in Figure 2.24, uses the For/Next looping structure to loop through the NameArray.

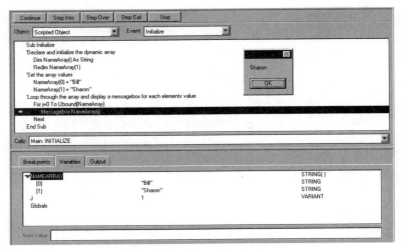

Figure 2.24   Looping through the NameArray

Erasing Array Values

You can use the Erase statement to remove

- All values, but keep the positions in a static array
- All values and elements in a dynamic array (this will recover the storage space until you use ReDim to rebuild the elements)

Syntax

```
Erase { arrayName | listName | listName ( tag ) }[, { arrayName |
listName | listName ( tag ) } ] . . .
```

Elements

arrayName   An array or a Variant variable containing an array. *array-Name* can end with empty parentheses.

listName   A list or a Variant variable containing a list. *listName* can end with empty parentheses.

tag   The list tag of a list element to be erased from the specified list.

Data Validation

Data Validation Functions

Several functions exist to determine the data type of a particular entity before operating on it, especially when dealing with variants. Some of these functions are

- IsDate
- IsNumeric
- Is Array
- Datatype

The Datatype function is useful in identifying the data type of an expression prior to testing it in other ways. Some of the more common return values are

- Zero is an empty string, and one is a null variant
- Two is an integer
- Eight is a string
- 35 is a product defined object such as a NotesDocument or NotesDatabase
- 8192 is a fixed array
- 8704 is a dynamic array

Check for Empty Variable

Depending on the data or object type you are working with, the test for an empty variable differs, as shown in Table 2.5.

Table 2.5 Tests for Empty Variables

Data Type	Test
Variant	IsEmpty to determine whether a variable is variant and has not yet been assigned a value originating in LotusScript (testing for a null string also works in most cases). Not applicable to testing Notes fields

continues

Table 2.5 Continued

Data Type	Test
Number	To test for an unassigned numeric data type originating in LotusScript, test the variable for a 0 value. (i.e. vNumber = 0). vNumber = " " to test for an empty Notes field
String	To test for an unassigned string value in LotusScript and Notes, test for a blank value (i.e. vString = " ")
Array	Without recursively accessing an array to check the value of each element, you can only test the highest (Ubound) or lowest (Lbound) boundary. Note that the elements may be allocated but not assigned

This code example, shown in Figure 2.25, uses some of the data validation functions and empty variable checks mentioned earlier.

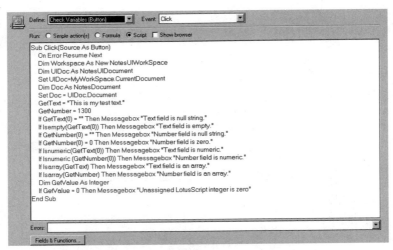

Figure 2.25 Data validation functions and empty variable checks

Check Empty Object

You can also check to see whether an object or object reference variable is empty (or invalid) by using Is Nothing, as shown in Table 2.6.

Table 2.6 Using is Nothing

Data Type	Test
Notes object	Use Is Nothing to test for the last item when recursively accessing Notes objects, such as documents in a view, ACL entries, or items in a document

Data Type	Test
Object reference variable	Use Is Nothing to test for an empty object handle to trap for the *Object Variable Not Set* error

The following example, shown in Figure 2.26, loops through the data directory (using the NotesDbDirectory object and GetFirstDatabase/GetNextDatabase methods) to display the entries, until the NotesDbDirectory object is Nothing (an empty product object).

Figure 2.26 Looping through the data directory

This example, shown in Figure 2.27, tests for a lack of an object handle while calling the NotesDatabase object Open method. ·

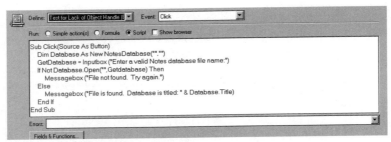

Figure 2.27 Testing for the lack of an object handle

Error Trapping

A run-time error occurs either when LotusScript executes an Error statement or when executing some other statement results in an error.

First, the error is recorded as the current error. LotusScript then records the line number in the script where the error occurred, the error number, and the error message associated with that number, if any. Until an error-handling routine is invoked for this error or another error is encountered, these are, respectively, the return values of the functions Erl, Err, and Error$. (Exception: The Err statement can be used to reset the current error number returned by the Err function.)

LotusScript then looks in the current procedure for an On Error statement associated with this error: either an On Error *n* statement, where *n* is the error number, or an On Error statement with no error number specified. If none is found, the search continues in the procedure that called this procedure, if any, and so on. If no associated On Error statement is found in any calling procedure, then execution ends—and the associated error message is displayed.

If an associated On Error statement is found, the following results can occur:

- If the On Error statement specifies Resume Next, execution continues in the procedure that contains the On Error statement. Execution continues with the statement following the statement that caused the error. (If the On Error statement is in a calling procedure, then the procedure call is considered to have caused the error—and all nested procedures are terminated before continuing execution.) The error is considered handled, but the return values of the functions Erl, Err, and Error$ are not reset.

- If the On Error statement specifies GoTo *label*, then execution continues at the statement labeled *label*. This event must be the same procedure as the On Error statement.

In this case, the first Resume statement encountered after the labeled statement ends the error processing. The error is considered handled when

- The return values of the functions Erl, Err, and Error$ are reset to their initial values: line number 0, error number 0, and the empty string (" ") as the error message, respectively.
- Execution continues at the location specified by the Resume statement.

In short, there are basically four choices when using the error-handler:

- When the error occurs, continue execution with the line following the error, or Goto a specific label to continue execution.
- Trap all errors and handle them in one generic manner, or trap for specific errors and handle each of them differently.

Figure 2.28 illustrates these choices.

	Trap All Errors	Trap Specific Errors
Continue with Next Line on Error	On Error Resume Next	On Error 99 Resume Next
Goto Label on Error	On Error Goto MyErrorFix	On Error 99 Goto MyErrorFix

Figure 2.28 Error-handler choices

All choices rely on the On Error statement, which once activated, is in effect from the time it runs until.

- The event procedure ends
- If in a sub-procedure (such as a user-defined function)
- The sub-procedure returns control to the calling script

On Error cannot be used at the form level to cover all object event errors, which means you must include it in every object event if you want error-handling to occur.

On Error Resume

The Resume function directs LotusScript to resume script execution at a particular statement in a script, after an error has occurred.

Syntax

```
Resume [ 0 | Next | label ]
```

Elements

0 Resumes execution at the statement that caused the current error.

Next Resumes execution at the statement following the statement that caused the current error.

label Resumes execution at the specified label.

You should use the Resume statement only in error-handling routines; once LotusScript executes the Resume statement, the error is considered handled.

Resume continues execution within the procedure where it resides. If the error occurred in a procedure called by the current procedure, and the called procedure did not handle the error, then Resume assumes that the statement calling that procedure caused the error:

- Resume [0] directs LotusScript to execute again the procedure-calling statement that produced the error.

- Note that this may result in an infinite loop, where in every iteration, the procedure generates the error and then is called again.

- Resume Next directs LotusScript to resume execution at the statement following the procedure call.

The Resume statement resets the values of the Err, Erl, and Error functions. An example is shown in Figure 2.29.

In this example, if you did not have the On Error statement, an error message would appear to the user indicating which error occurred completely—stopping the script from executing. By using Resume Next, the rest of the script completes. Note that this could cause problems if an error occurs in a critical line, on which subsequent lines are dependent. In this example, if more calculations subsequent to the error (which occurs when calculating the Result variable) relied on the value of the Result variable, more errors would occur—and your expected result may not happen.

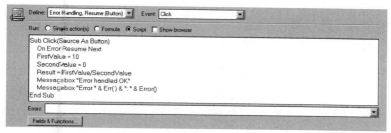

Figure 2.29 The Resume statement

On Error Goto

When using On Error Goto when an error occurs, control of the procedure is passed to the label specified. This feature enables you to create a routine specifically for error-handling. This routine can handle the errors in a generic fashion, or it can handle specific errors.

Syntax

```
On Error [ errNumber ] { GoTo label }
```

Elements

errNumber   Optional. An expression whose value is an Integer error number. If this is omitted, this statement refers to all errors in the current procedure. This value can be any error number that is defined in LotusScript at the time the On Error statement is encountered.

GoTo label   Specifies that when the error *errNumber* occurs, execution continues with an error-handling routine that begins at *label*. The error is considered handled.

Figure 2.30 shows an example.

The error-handler in this code, in this case, constitutes the last lines in the script. After the message box displays, the Sub could have ended—but in this case, it returns to the line that caused the problem. The error-handler includes a routine to fix the problem by asking the user to re-enter a filename.

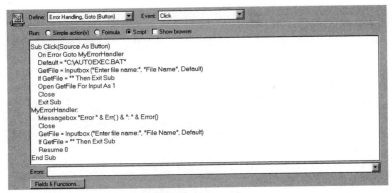

Figure 2.30 Using On Error Goto

Specific Error Trapping

Every error recognized at run-time has its own error number that identifies it. When a recognized error happens during script execution, LotusScript records the error number and then proceeds as directed by an On Error statement that refers to that number.

For example, you might write either one of these On Error statements to tell LotusScript how to respond to an occurrence of error number 357:

```
On Error 357 GoTo MyErrorHandler
On Error 357 Resume Next
```

Error numbers are established in two ways:

- By predefinition in LotusScript. LotusScript recognizes many common errors and has a built-in error number associated with each one. The text file LSERR.LSS defines constants for each error. The value of the constant is the error number. To make these available to your script, include this file in your script with the statement:

```
%Include "LSERR.LSS"
```

- By an Error statement in a script. This statement signals error number 357

```
Error 357
```

When this statement is executed, LotusScript records the occurrence of error 357 and then proceeds as directed by an On Error 357 statement. This facility has two uses:

1. You can use it to signal an error, give the error a number, and trigger error processing for that error. This feature is how you augment the predefined errors with errors and error processing specific to the needs of your script.

2. You can use it to simulate a predefined error. This command is how you force LotusScript to execute some error-processing code, without depending on the error to occur while other statements are executing. This feature is useful for testing the error-processing logic in your script.

When referring in an On Error statement to a predefined error, you can use the constant for the error, not the error number itself. For example, here are the statements in LSERR.LSS that define the error numbers and constants for two common errors:

```
Public Const ErrDivisionByZero     = 11 ' Division by zero
Public Const ErrIllegalFunctionCall = 5  ' Illegal function call
```

On Error statements then do not need to mention the numbers 11 and five. Write the statements in the following form instead, making the script easier to read.

```
On Error ErrDivisionByZero  . . .
On Error ErrIllegalFunctionCall  . . .
```

The error constants are defined in the following files in the Notes data directory. You must include the appropriate file or files in the (Declarations) event of the object to access the error constants. You can create a script library with the %include statements, and then use that script library in the options section.

- lsxbeerr.lss defines constants for errors raised by Notes back-end methods.

- lsxuierr.lss defines constants for errors raised by Notes front-end (UI) methods.

- lserr.lss defines constants for errors raised by LotusScript.

- lsconst.lss defines constants for use as arguments in LotusScript statements and functions.

Similarly, you can define constants for your own error numbers. Then the constant names can be used instead of numbers in any Error statements and On Error statements that refer to the error. For example

```
Const ooBounds = 677          ' A specific out-of-bounds error
On Error ooBounds…
```

Figure 2.31 shows an example.

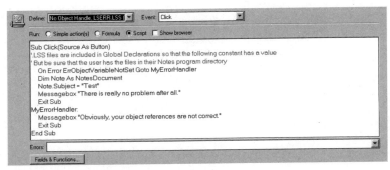

Figure 2.31   Defining constants

User-Defined Procedures

Custom Procedures

LotusScript defines a host of functions that you can call from predefined object procedures. You can also create your own procedures that enable you to *reuse* code, without having to repeat the same code over and over in various object events. You can call the procedures (and pass them parameters) just like other built-in LotusScript functions.

Procedures can be either Sub or Function, with both executing lines of code that reside outside the object event procedure execution. The primary difference between the two types of procedures is that a Function returns a value, and a sub does not.

Advantages of User-Defined Procedures

Though technically slower than having all the code *in-line*, the advantages to creating and using your own procedures is

- Code is more modular, and, therefore, is easier to troubleshoot.
- Overall script size is smaller, because the code is reused.
- Procedures can be placed in a .LSS file, and using the %Include c:\notes\mylib.lss directive (in the (Global) (Declarations) event area, they can be called by any application (though this means that every user must have the .LSS file in a specified directory).

Where to Define Procedures

Procedures are typically placed in the (Globals) event area of a form and are created by typing over any existing procedure with the new procedure name. For example, navigate to the (Globals) Initialize event, shown in Figure 2.32, which displays the following items.

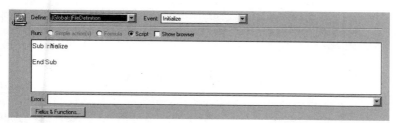

Figure 2.32 The (Globals) Initialize event

Highlight all the lines (Ctrl-A) and type in a procedure name. By typing over this sub and typing in a new procedure name, LotusScript will automatically create the new procedure.

Note that you must include the statement Option Public in the (Global)(Options) event area, or the procedure will not be available to other object events in the form.

Figure 2.33 shows a Sub Procedure example.

This example shows a Sub that, when called while a document is in Edit mode, puts the focus on a specified field. The Sub expects a parameter to be passed to it. To call this user-defined sub and pass it a parameter, you would write:

```
Call SetFieldFocus("Amount")
```

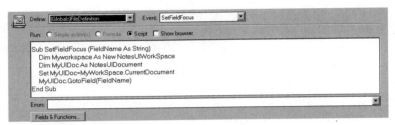

Figure 2.33   Sub Procedure example

The string *Amount* in our example is a field name on a form. The sub refers to that value as FieldName and passes it to the NotesUIDocument object method, GotoField.

If you use this function in more than one field Exiting event on a form, you may create a *nested form*. This loop is a type of loop that results when the exiting of one field triggers both the Exiting event of the field and the Entering event of the next field. When the user moves from the second field, its Exiting event runs.

Figure 2.34 shows a Function Procedure example.

Figure 2.34   Function Procedure example

This example shows a function that, when passed a value, converts it to another value and returns it to the calling sub.

The following example calls the function, as shown in Figure 2.35.

In this example, the function is called in the last line of the Sub. The ConvertTemp function returns the value to the message box statement that displays it.

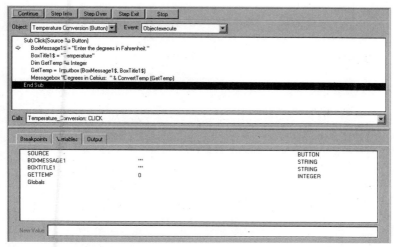

Figure 2.35 Calling the function

Sample Questions

1. *Objective:* Variable Declaration

   When implicitly declaring a string variable, the proper syntax is

   a. Dim vString As String
   b. Dim vString As String$
   c. Dim vString%
   d. Dim vString$

2. *Objective:* Literal String Construction Rules

   A literal string in LotusScript is a string of characters enclosed in the following sets of delimiters:

   a. Quotes (" ")
   b. Brackets ([ ])
   c. Quotes (" ") and Brackets ([ ])
   d. Braces ({ }), Bars (||) and Quotes (" ")

3. *Objective:* Statements and Functions

   One difference between Statements and Functions is

   a. There is no difference.
   b. Statements return values, and Functions do not
   c. Functions return a value, and Statements do not
   d. Statements and Functions both return values

4. *Objective:* Control Structures

   If you do not use a branching structure

   a. A script will execute line-by-line in sequence.
   b. A script will execute Subroutines first.
   c. A script will not execute without a branching structure.
   d. A script will execute line-by-line from the bottom-up.

5. *Objective:* Arrays

A dynamic array

a. Must have its element defined explicitly
b. Can have more than eight dimensions
c. Does not hold any places for its elements in memory when first declared
d. Holds a place in memory for its elements when first declared

6. *Objective:* Referencing Array Values

When referencing an array value

a. You must loop through the entire array until you get to the value you are looking for.
b. You must reference the value by its numeric position in the array.
c. The array value must be referenced by using the Ubound function.
d. You must reference the value by using a for/next looping structure.

7. *Objective:* Error-Handling

Once an On Error statement is activated

a. The event procedure ends.
b. If in a sub-procedure (such as a user-defined function)
c. Until the sub-procedure returns control to the calling script
d. All of the above

8. *Objective:* User-Defined Procedures

One of the advantages of user-defined procedures is

a. Overall script size is smaller, as the code is reused
b. A user-defined procedure is technically more efficient.
c. They are the only place where branching structures are used.
d. There are no advantages to using a user-defined procedure.

9. *Objective:* Retrieving and Assigning Arrays Values

   Which line will cause an error?

   ```
   EmployeeNames = Doc.NameArray-Line 1
   Department = Doc.DepartmentArray
   Forall n In EmployeeNames
   If n = "Mark Ploch" Then-Line 2
   Redim Preserve Department(Ubound(Department)+1)-Line 3
   Department(Ubound(Department) + 1) = "Acquisitions"-Line 4
   End If
   End Forall
   ```

   a. Line 1

   b. Line 2

   c. Line 3

   d. Line 4

10. *Objective:* Retrieving and Assigning Arrays Values

    Jim is trying to loop through the company employees and put up a message box that will inform each department. Which line will cause an error?

    ```
    EmployeeNames = Doc.NameArray
    Department = Doc.Department-Line 1
    For j = 0 To Ubound(EmployeeNames)-Line 2
    If EmployeeNames = "Mark Ploch" Then-Line 3
    Messagebox "Mark works for the " & Department(0) & "
    department."-Line 4
    End If
    Next
    ```

    a. Line 1

    b. Line 2

    c. Line 3

    d. Line 4

Sample Answers

1. *Answer:* d

   When declaring variables implicitly, you do not need to explicitly name the data type. The data type suffix for a string is $. The % suffix is for an integer.

2. *Answer:* d

3. *Answer:* c

   Functions perform an action and return a value to a declared variable or a variant.

4. *Answer:* a

   If there are no branching structures in the script, it will execute each line in sequence.

5. *Answer:* c

   A dynamic array will not hold a place in memory for its elements until it is ReDimmed and its number of elements is defined. No array can have more than eight dimensions, whether static or dynamic.

6. *Answer:* b

   When referencing an array, you must reference the value by its numeric position in the array.

7. *Answer:* d

8. *Answer:* a

   The main benefits of using user-defined procedures are

   - Code is more modular, and therefore is easier to troubleshoot.
   - Overall script size is smaller, because the code is reused.
   - Procedures can be placed in a .LSS file using the %Include `c:\notes\mylib.lss` directive.

9. *Answer:* d

```
EmployeeNames = Doc.NameArray
Department = Doc.DepartmentArray
Forall n In EmployeeNames - Line 1
If n = "Mark Ploch" Then - Line 2
Redim Preserve Department(Ubound(Department)+1) - Line 3
Department(Ubound(Department) + 1) = "Acquisitions" - Line 4
End If
End Forall
```

Line 4 causes the error because it is trying to set a non-existent array element. We increase the number of elements in the Department array in Line 3. In Line 4, we get the upper bounds of the Department array, and then add one to that number and try and set it. This event causes a *Subscript out of range* error. The line should read

```
Department(Ubound(Department)) = "Acquisitions"
```

10. *Answer:* c

```
EmployeeNames = Doc.NameArray
Department = Doc.Department - Line 1
For j = 0 To Ubound(EmployeeNames) - Line 2
If EmployeeNames = "Mark Ploch" Then - Line 3
Messagebox "Mark works for the " & Department(0) & "
department." - Line 4
End If
Next
```

Line 3 incorrectly references the EmployeeNames array. What LotusScript is looking for is which element of the array it should be comparing with *Mark Ploch*. In other words, there is no array element reference. The line should read

```
If EmployeeNames(j) = "Mark Ploch" Then...
```

The reason for the j (as opposed to a hard-coded number) is because we are looping through the array using a For/Next looping structure. As we loop, j will increment by one until it reaches the upper bounds of the EmployeeNames array.

CHAPTER 3

Scripting Lotus Notes Objects

This chapter introduces you to the concepts of object orientation and what it means with respect to Lotus Notes and LotusScript. You will learn how to create objects and access their properties and methods. This chapter will also introduce you to the concept of reusing code by taking advantage of a script's scope.

TIP

Again, this chapter covers the basic building blocks of coding with Lotus Script, in relation to objects, their creation and property and method access. On the exam, you will be asked to debug code—and if you are not familiar with these concepts, you will be forced to guess between two similar answers.

Chapter Objectives

The objectives of this chapter are to increase your understanding of the following items:

- LotusScript object orientation
- Creation of Lotus Notes objects
- Object properties

- Object methods
- Script scope and source issues

Object Orientation

Though the LotusScript language primitives are common across all products, each product includes its own unique object classes that you can use to accomplish the goals of the application. Classes are common to object-oriented programming and are used to represent objects whose data can be protected, initialized, and accessed by a specific set of procedures (properties and methods). A class enables your application to model real objects, their attributes, and their behaviors. Within Notes, for example, there are numerous classes: NotesDatabase, NotesDocument, NotesAgent, and NotesUIWorkspace classes, for example.

Each class contains a series of properties and methods. For example, the NotesDatabase class contains a DatabaseTitle property that returns or sets the database title—and a Replicate method with starts replication for a database. All of the Notes product objects are outlined in an object model in a hierarchical manner. Some objects are containers for other objects. A database object, for example, is a container for document and view objects (among others). To take this out of the context of programming, the object model makes logical sense. If we look at a Notes database in pieces, a database contains various design elements such as forms, views and agents. Not only does it contain design elements, but it also contains data in the form of Notes documents. If we think about each of these as containers, literally, when we open a database, we have access to views, forms and agents. If we open a view container, we have access to Notes documents or data. If we open a Notes document, we then have access to NotesItems (or fields). The object model, in essence, is a map which shows you how to get from one place to another using Notes object classes.

Objects

You use object reference variables to create, manage, and delete objects. An object reference variable is different from other variables, because it is associated with an instance of a class (that is, an object). The object reference variable, like other variables, has a named area in storage. Unlike other variables, however, the value stored in the area named by an object reference variable is not the object itself. Instead, the object (and the data it consists of) is stored elsewhere. The value stored in an object reference variable is a 4-byte pointer to the object data, called the object reference. LotusScript uses this pointer to access the object data. When you assign a value to an object reference variable, you associate the object reference with the object. For example, when we declare an object reference variable for a NotesDatabase, all of the data associated with the NotesDatabase object is not stored in memory. Instead, when we use the object reference variable, it points to the database where the data we need is stored.

Creating Objects

When you create an instance of a class, you must explicitly declare an object reference variable. In other words, you create the object, create the object reference variable, and assign an object reference to the variable. The object reference points to the object. When an object is created, its member variables are initialized, each to the initial value for the data type of the member. For example, a member of data type Integer is initialized to zero. You can create an object reference—without creating an object—with the following syntax:

```
Dim x As ClassName
```

Because the variable you declare contains a reference to an object that does not yet exist, the variable is initialized to the value of NOTHING.

You create objects using the LotusScript New keyword. You can use the New keyword with either the LotusScript Dim or Set statement. You can use the keyword New in the declaration that declares an object reference variable. If you use the New keyword when you declare the object reference variable, the declaration creates an object and assigns to the variable a reference to the newly created object.

- To create a new object and assign a reference to that object in a variable that you are declaring, use the Dim statement with the following syntax:

```
Dim objRef As New className[(arglList)]
```

- To create a new object and assign reference to it if you have already declared an object reference variable (with a Dim statement without the New keyword), use the Set statement with the following syntax:

```
Set objRef = New className[(argList)]
```

You are not using the New keyword to declare an array of object reference variables or a list of object reference variables.

Figure 3.1 shows an object declarations example.

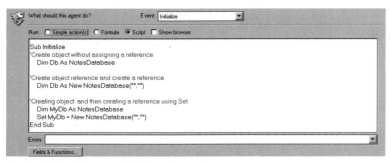

Figure 3.1 Object declarations example

You can create an instance of a class by using a Set statement that includes the New keyword and a variable that was previously declared as an object reference variable for that class, as in the earlier example.

The Set statement is a kind of assignment statement used only to assign values (object references) to object reference variables. You cannot use it to assign values to any other kind of variable.

In summary, every Notes object declared and used in a script represents an instance of a particular class of object.

Every item (field) created on a form, for example, is an instance of a product-defined object that inherits its behavior from the NotesItem class.

The object class is a blueprint, or road map, which defines what operations the object, or run-time instance of an object class, can perform.

Your interaction with Notes objects involves three steps:

1. Declaring an object reference variable

2. Setting the object reference variable as an instance of the object class

3. Manipulating the object reference variable to call the methods and properties of the object using script

Object Class Overview

Table 3.1 provides a brief overview of some common Lotus Notes product-specific objects. Notes defines LotusScript classes that enable you to access Notes structures at two levels:

- The database (back-end) classes enable you to access named databases, views, documents, and other Notes objects. Both workstation and server users can run scripts that access database objects.

- The UI (front-end) classes enable you to access current objects being worked on by the user. Only workstation users can run scripts that access UI objects.

Table 3.1   Back End Classes

Class	Description
NotesACL	Represents a collection of all the access control list entries for a database
NotesACLEntry	Represents a single entry in an access control list
NotesAgent	Represents an agent
NotesDatabase	Represents a Notes database
NotesDateRange	Represents a range of dates and times
NotesDateTime	Provides a means to translate between LotusScript and Notes date/time formatting
NotesDbDirectory	Represents the database files on a server or the local machine
NotesDocument	Represents a document in a database

continues

Table 3.1 Continued

Class	Description
NotesDocumentCollection	Represents a collection of documents
NotesEmbeddedObject	Represents embedded objects, links, and file attachments
NotesForm	Represents a form in a database
NotesInternational	Represents the international settings in the operating system
NotesItem	Represents a piece of data in a document
NotesLog	Represents actions and errors that occur during a script's execution
NotesName	Represents a user or server name
NotesNewsLetter	A summary document that contains information from, or links to, several other documents
NotesRichTextItem	Represents items that can contain rich text
NotesSession	Root of Notes database objects—for global attributes, context, and persistent information
NotesTimer	A mechanism for triggering an event every fixed number of seconds
NotesView	Represents a named view of a database
NotesViewColumn	Represents a column of a view

Front End Classes

Table 3.2 shows some front-end classes.

Table 3.2 Front End Classes

Class	Description
Button	Represents an action, button, or hotspot on a form or document
Field	Represents a field on a form
Navigator	Represents an object in a navigator
NotesUIDatabase	Represents the database that is currently open in the Notes workspace
NotesUIDocument	Models the behavior of a Notes document window
NotesUIView	Represents the current database view
NotesUIWorkspace	Provides access to the current workspace

Object Container Example

This object container example illustrates the object model *road map* and how container objects work. In Table 3.3, we follow the NotesDatabase object to the NotesItem object using container objects. As with a *road map*, there are many different ways to get from one place to another. This method is only one way to get from the NotesDatabase object to the NotesItem object. There are other ways to do this as well.

Table 3.3 Object Container Examples

Object	Script
NotesDatabase	Dim Db As New NotesDatabase(" "," ") Call Db.Open(" ","MyDatabase.NSF")
NotesView (contained by NotesDatabase)	Dim Db As New NotesDatabase(" "," ") Call Db.Open(" ","MyDatabase.NSF") Dim View As NotesView Set View = Db.GetView("AllNames")
NotesDocument (contained by NotesView)	Dim Db As New NotesDatabase(" "," ") Call Db.Open(" ","MyDatabase.NSF") Dim View As NotesView Set View = Db.GetView("AllNames") Dim Doc As NotesDocument Set Doc = MyView.GetFirstDocument
NotesItem (contained by NotesDocument)	Dim Db As New NotesDatabase(" "," ") Call Db.Open(" ","MyDatabase.NSF") Dim View As NotesView Set View = Db.GetView("AllNames") Dim Doc As NotesDocument| Set Doc = MyView.GetFirstDocument Dim Item As NotesItem Set Item = Doc.GetFirstItem("Name") Messagebox Item.Value(0)

Object Property Access

Notes object instances have predefined properties and behaviors that you cannot change programmatically, such as how an editable field has a default, translation, and validation formula, or the font setting for a Text field on a form.

Most objects have

- Properties that you can get or set using LotusScript, such as the value of a field

- Behaviors, or methods, and subprograms that you can invoke, such as when saving a document or creating and populating an item (field)

From this object-oriented perspective, programming with Lotus-Script consists of creating instances of product objects—and controlling their properties and methods to build the logic to control the application.

Properties

Object properties are the adjectives, or descriptors, of objects. The most obvious property of an object is its value, which may be set by a user or set programmatically by a script.

Notes product objects have predefined properties. For example, an instance of the NotesDatabase class object may be described by its filename, title, replica ID, and last modification date, just to name a few.

Some properties include the following items

- *Read-only.* You can only retrieve a value that was determined previously by another Notes process. The filename of a database object is an example of a read-only property.

- *Read/write.* The database title, for example, can be changed programmatically using the NotesDatabase object title property.

Accessing Properties

The properties of an object can be read or set once the object is declared and is given a *handle*, or object variable reference, as shown in Figure 3.2. For example, to access the common name of the user, you must first declare the NotesSession object. Then using the dotted notation <object>.<property>, access the property.

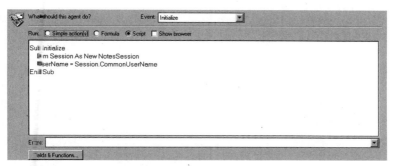

Figure 3.2 Giving an object a handle

When working with objects, you must associate the instance of the object with an object label, or *object handle*, which enables you to distinguish it from other instances of the same class (such as multiple field objects in a document).

In the previous example, *Dim Session As New NotesSession* declares an object variable called Session as an instance of the NotesSession class. From that point on, you must refer to that specific instance using the variable name Session.

Many properties return an array of values that must be referenced using the element subscript number. You can loop through these properties using a For/Next loop. For example, the NotesView property Columns returns an array of all the columns in the view, as shown in Figure 3.3.

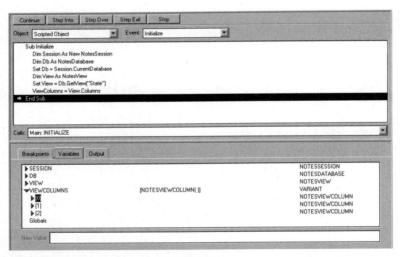

Figure 3.3 NotesView property Columns

In this example, the ViewColumns variant holds an array of NotesViewColumn objects. This array was created by using the NotesView Columns property. Figure 3.4 shows the NotesViewColumn properties of one of the NotesViewColumn elements within the ViewColumns array.

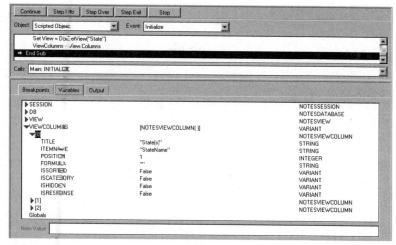

Figure 3.4 NotesViewColumn properties

Object Method Access

Object methods are the verbs of objects. In other words, they are predefined subprograms that are executed by (and on) the objects.

As with properties, Notes product methods have predefined behaviors for each type of object. You have seen how the GetView method of the NotesDatabase object class is used to define an object variable for a specific view. Using the object variable to represent the view object, further methods can be called from the NotesView object, as seen in Figure 3.5.

Figure 3.5 Calling methods from the NotesView object

Using Methods

The methods of an object can be executed once the object is declared and given a handle, or object variable reference. For example, to execute the Open method of the NotesDatabase object, you must first declare the NotesDatabase object. Then, using the dotted notation <object>.<method>, execute the method.

Once the database object is declared and set, you can declare the NotesView object. Using the NotesDatabase object GetView method, set the NotesView object to a particular view.

- Methods that set an object variable are executed using the Set syntax.

- Methods that perform an action (but do not set an object variable) are executed using the Call syntax. They typically return a value to a variable as well.

In Figure 3.6, the NotesDatabase method Compact returns a value to the amtShrunk variable:

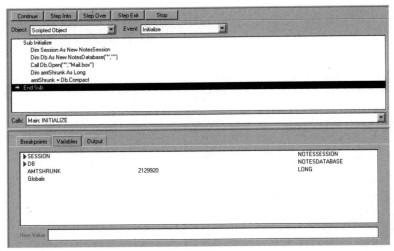

Figure 3.6 NotesDatabase method Compact

Reusing Objects

To produce the greatest flexibility of code from various design elements, and to make the code more legible and understandable for yourself and for other programmers, it is good practice to modularize your code to achieve the following effects:

- Centralizing all declarations of variables and object reference variables in the (Globals)(Declarations) event
- Using the QueryOpen or PostOpen events to set object variables used in field object methods
- Reusing the same object variable names by declaring them once and setting them multiple times

Although more transportable and easier to understand at first, storing all declarations and Sets together inside one specific event is

- Less efficient, in that you are writing, compiling, and running redundant code
- Less versatile, in that you are not reusing centrally stored code in multiple events

Reusing objects uses the same principle as declaring user-defined procedures: you declare the object once, and then you use it anywhere in the form. Publicly declaring objects is optional if you do not intend to re-access the same object instance. In many other cases, you can go either way; but in other cases, it is absolutely essential, such as during iteration through an object array using GetNext. If you reset the object, the pointer returns to the *top of the object stack.*

Scope of Declarations in LotusScript

Scope is the context in which a variable, procedure, class, or type is declared. Scope affects the accessibility of an item's value outside the context in which it was declared. For example, variables declared within a procedure are typically not available outside the scope of that procedure.

LotusScript recognizes three kinds of scopes:

- Module scope
- Procedure scope
- Type or class scope

 NOTE
It is important to understand these concepts for the exam. You will be asked questions that test your knowledge on this subject.

Name Conflicts and Shadowing

Two variables or procedures with the same name cannot be declared in the same scope. The result is a *name conflict*. The compiler reports an error when it encounters a name conflict in a script.

Variables or procedures declared in different scopes can have the same name; however, when such a name is used in a reference, LotusScript interprets it as referring to the variable or procedure declared in the innermost scope that is visible where the reference is used.

A variable or procedure of the same name, declared at a scope outside of this innermost visible scope, is not accessible. This effect is called shadowing. The outer declaration(s) of the name are *shadowed,* or made invisible, by the inner declaration.

Module Scope

A variable is declared in module scope if the declaration is outside any procedure, class, or type definition in the module. The variable name has a meaning as long as the module is loaded.

The variable name is visible anywhere within the module. Everywhere within the module, it has the meaning specified in the declaration—except within a procedure, type, or class where the same variable name is also declared.

The variable is Private by default and can be referred to only within the module that defines it. A variable can be referred to in other modules only if it is declared Public—and the other modules access the defining module with the Use statement.

The following situations result in a name conflict across modules:

- Two Public constants, variables, procedures, types, or classes with the same name
- A Public type with the same name as a Public class
- A Public module-level variable with the same name as a Public module-level constant or procedure
- A Public module-level constant with the same name as a Public module-level procedure

The following situations result in a name conflict within a module:

- A type with the same name as a class
- A module-level variable with the same name as a module-level constant or procedure

- A module-level constant with the same name as a module-level procedure

Procedure Scope

A variable is declared in procedure scope if it is declared within the definition of a function, sub, or property. Only inside the procedure does the variable name have the meaning specified in the declaration. The variable name is visible anywhere within the procedure.

Ordinarily, the variable is created and initialized when the procedure is invoked, and it is deleted when the procedure exits. This behavior can be modified with the Static keyword as follows:

- If the variable is declared with the Static keyword, its value persists between calls to the procedure. The value is valid as long as the module containing the procedure is loaded.

- If the procedure itself is declared Static, the values of all variables in the procedure (whether explicitly or implicitly declared) persist between calls to the procedure.

The following situations result in a name conflict within a procedure.

- Two procedure arguments with the same name
- Two labels with the same name
- Two variables with the same name
- A procedure argument and a variable with the same name
- A function that contains a variable or argument of the function name
- A property that contains a variable of the property name

Type or Class Scope

A variable is declared in type or class scope if it is declared within the definition of a type or a class. For classes, it must additionally be declared outside the definition of a procedure. The variable is called a member variable of the type or class.

Type member variables: A type member variable is created and initialized when an instance of that type is declared—and is deleted when the type instance or instance variable goes out of scope. The visibility of a type member variable is automatically Public.

Class member variables: A class member variable is created and initialized when an instance of that class is created, and it is deleted when the object is deleted.

Each class member variable can be declared Public or Private. A Private member can only be referred to within the class or its derived classes; however, class member variables are Private by default.

The visibility of a type member variable (which is always Public) and a Public class member variable depends, for any particular type or object, on the declaration of the instance variable that refers to that instance.

- If the instance variable is declared Private, then the member variable is visible only in the owning module.

- If the instance variable is declared Public, then the member variable is visible wherever the instance variable is visible. It can be referred to in the other modules where the module that owns this instance variable is accessed with the Use statement. A name conflict within a type occurs with two type members with the same name, whereas a name conflict within a class occurs when two class members (variables or procedures) have the same name.

Global Declarations

You can declare an object variable, instantiate it at the global level, and use it across the form, as shown in Figures 3.7, 3.8, 3.9, and 3.10. The NotesSession object is instantiated at the global level and is used in a button on the form. Notice that there is no need to declare the NotesSession object reference in the button code (see Figure 3.10).

- (Globals)(Options)

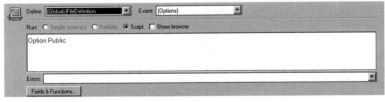

Figure 3.7 Set options.

- (Globals)(Declarations)

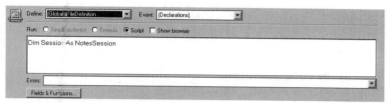

Figure 3.8   Declare a NotesSession.

- (Form)(QueryOpen)

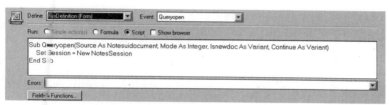

Figure 3.9   Set the NotesSession.

- Prompt Name button code (Click) event

Figure 3.10   Use the session object in (Globals).

Now that the Session object reference variable is set, it can now be accessed in any design element on the form.

Resetting and Reusing Objects

You can reset an object variable that has been set at the global level and set/reset it locally, as seen in Figures 3.11, 3.12, 3.13, and 3.14.

You can also use the same object instantiation in different parts of the form. In this example, the Db object reference variable is reset multiple times and is reused in the Reset Database button.

- (Globals)(Options)

Figure 3.11 Set options.

- (Globals)(Declarations)

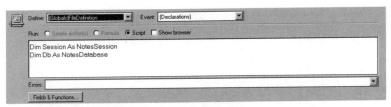

Figure 3.12 Declare the NotesSession and NotesDatabase.

- (Form)(QueryOpen)

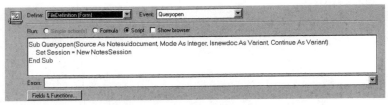

Figure 3.13 Set the NotesSession object.

- Reset Database button (Click) event

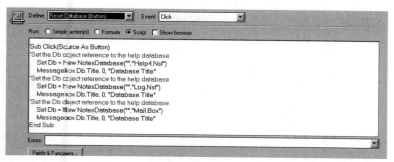

Figure 3.14 Reuse and Reset the NotesDatabase object declared in (Globals).

Sample Questions

1. *Objective:* Object orientation

   Classes are used

   a. To understand Notes objects
   b. To declare methods
   c. To represent objects whose data can be protected, initialized, and accessed by properties and methods
   d. Only in Lotus Notes

2. *Objective:* Objects

   An object reference variable is different from other variables because

   a. It stores its values in memory.
   b. It is associated with an instance of a class.
   c. There is no difference.
   d. It can only be used with certain data types.

3. *Objective:* Creating objects

   When you create an instance of a class

   a. You can use a data type suffix.
   b. You use the syntax Dim notesClass As x.
   c. You use the syntax Dim x As notesClass.
   d. You cannot assign an object reference later.

4. *Objective:* Creating objects

   You create objects using the LotusScript keyword

   a. Set
   b. DefType
   c. New
   d. Call

5. *Objective:* Object class overview

   LotusScript classes enable you to access Notes structures at which two levels?

   a. Database and view

   b. Database and back-end

   c. Front-end and view

   d. Front-end and back-end

6. *Objective:* Object class overview

   An example of a front-end class is

   a. NotesDatabase

   b. NotesDocument

   c. NotesUIView

   d. NotesView

7. *Objective:* Object access

   Most objects have

   a. Properties and methods

   b. Properties only

   c. Methods only

   d. None of the above

8. *Objective:* Accessing methods

   When accessing an object's methods, the correct syntax to use is

   a. <object>.<database>

   b. <object>.<method>

   c. <method>.<object>

   d. <database>.<property>

9. *Objective:* Reusing objects

   Storing all declarations and Sets together inside one specific event

   a. Is very efficient
   b. Enables you to reuse code
   c. Is less versatile in that you are not reusing centrally stored code in multiple events
   d. Is not important

10. *Objective:* Scope of declarations

   LotusScript recognizes three types of scopes, one of which is

   a. Module
   b. Procedure
   c. Type or class
   d. All of the above

Sample Answers

1. *Answer:* c

   Classes are common to object-oriented programming and are used to represent objects whose data can be protected, initialized, and accessed by a specific set of procedures (properties and methods).

2. *Answer:* b

   An object reference variable is different from other variables, because it is associated with an instance of a class (that is, an object). Most other variables are declared a particular data type and do not reference any classes.

3. *Answer:* c

   When you create an instance of a class, you must explicitly declare an object reference variable. There are no implicit declarations of classes.

4. *Answer:* c

   You create objects using the LotusScript New keyword: Dim *x* As *New NotesUIWorkspace.*

5. *Answer:* d

   LotusScript classes enable you to access Notes structures at two levels:

   - The database (back-end) classes enable you to access named databases, views, documents, and other Notes objects. Both workstation and server users can run scripts that access database objects.

   - The UI (front-end) classes enable you to access current objects being worked on by the user. Only workstation users can run scripts that access UI objects.

6. *Answer:* c

   Any object with UI (User-Interface) in the name usually corresponds to a front-end class. Not all front-end classes have UI in their name, however.

7. *Answer:* a

Most objects have

- Properties that you can get or set using LotusScript, such as the value of a field
- Behaviors, or methods, and subprograms that you can invoke, such as saving a document or creating and populating an item (field)

8. *Answer:* b

The methods of an object can be executed once the object is declared and is given a handle, or object variable reference. You must first declare the Notes object. Then, using the dotted notation <object>.<method>, execute the method.

9. *Answer:* b

Although more transportable and easier to understand at first, storing all declarations and Sets together inside one specific event is

- Less efficient, in that you are writing, compiling, and running redundant code
- Less versatile, in that you are not reusing centrally stored code in multiple events

10. *Answer:* d

LotusScript recognizes three kinds of scopes:

- Module scope
- Procedure scope
- Type or class scope

CHAPTER 4

Database Access

With this chapter, we begin to dig into the actual Lotus-Script object classes. This chapter reviews the NotesDatabase class and some of its properties and methods.

TIP
It is important to know the syntax for creating a new database, creating a replica copy of a database, deleting a database, and opening a database. You will be expected to catch syntax errors on the exam.

Chapter Objectives

The objectives of this chapter are to increase your understanding of

- Establishing connectivity within a database
- Establishing connectivity between Notes databases
- Creating a database
- Deleting a database
- Creating a new replica database
- Copying a database
- Accessing a database's properties and methods

NotesDatabase Class

The NotesDatabase class provides a means for locating and opening Notes databases. A NotesDatabase object provides access to NotesView, DocumentCollection, and NotesDocument objects. NotesDocument objects can be accessed directly through a NotesDatabase object, or by first obtaining a NotesDocumentCollection or NotesView object. A NotesDatabase object also provides access to NotesACL and NotesForm objects.

Creation and Access

There are several ways a script can use the NotesDatabase class to access existing databases and to create new ones.

- To access an existing database when you know its server and filename, use New or the GetDatabase method in NotesSession.

- To access the database in which a script is currently running, without indicating a server or filename, use the CurrentDatabase property in NotesSession.

- To access an existing database when you know its server and replica ID, use the OpenByReplicaID method.

- To access an existing database when you know its server name but *not* its filename, use the NotesDbDirectory class.

- To access the current user's mail database, use the OpenMail method.

- To open the default Web navigator database, use the OpenURLDb method.

- To access the available Address books, use the AddressBooks property in NotesSession.

- To test for the existence of a database with a specific server and filename before accessing it, use one of these properties or methods: IsOpen, Open, or OpenIfModified.

- To create a new database from an existing database, use one of these methods: CreateCopy, CreateFromTemplate, or CreateReplica.

- To create a new database from scratch, use the Create method.

- To access a database when you have a NotesView, NotesDocument, NotesDocumentCollection, NotesACL, or NotesAgent from that database, use the appropriate Parent (or ParentDatabase) property.

TROUBLESHOOTING TIP
A database must be open before a script can use the properties and methods in the corresponding NotesDatabase object. In most cases, the class library automatically opens a database for you. But watch out for the following cases:

- A script that runs on a server cannot open a database that is on a different server. An error is returned if your script attempts to do so.
- A script that attempts to open a database to which it does not have access returns an error. A script needs at least Reader access to a database in order to open it.
- A NotesDatabase retrieved from a NotesDbDirectory object is closed. The following properties are available on the closed database: FileName, FilePath, IsOpen, LastModified, Parent, ReplicaID, Server, Size, SizeQuota, and Title. To access all the properties and methods of a database retrieved from a NotesDbDirectory, a script must explicitly open the database.
- A NotesDatabase retrieved from the AddressBooks property in NotesSession is closed. The following properties are available on the closed database: FileName, FilePath, IsOpen, IsPrivateAddress-Book, IsPublicAddressBook, Parent, and Server. To access all the properties and methods of a database retrieved from the Address-Books property, a script must explicitly open the database.
- A NotesDatabase retrieved using New is closed if no database exists at the server$ and dbfile$ specified. The following properties are available on the closed database: FileName, FilePath, IsOpen, Parent, and Server.

Notes returns an error when a script attempts to perform an operation for which it does not have appropriate access. The properties and methods that a script can successfully use on a NotesDatabase object are determined by these factors:

- The script's access level to the database, as determined by the database *Access Control List* (ACL). The ACL determines whether the script can open a database, add documents to it, remove documents from it, modify the ACL, and so on.
- The script's access level to the server on which the database resides, as determined by the Server document in the Public Address Book. The Server document determines when an agent can run and which LotusScript features it can use.

When a script runs on a server, the script's access level to databases and servers corresponds to the access level of the script's owner (the person who last saved the script). When a script runs on a workstation, the script's access level to databases and servers corresponds to the access level of the current user.

Syntax

```
Dim variableName As New NotesDatabase( server$, dbfile$ )
```

or

```
Set notesDatabase = New NotesDatabase( server$, dbfile$ )
```

The New keyword creates a NotesDatabase object that represents the database located at the server and filename you specify—and opens the database if possible. Unlike the behavior of New in other classes (such as NotesDocument), using New for a NotesDatabase does *not* create a new database on disk.

Parameters

server$ **String.** The name of the server on which the database resides. Use an empty string (" ") to indicate a database on the current computer. If the script runs on the workstation, the empty string indicates a local database. If the script runs on a server, it indicates a database on that server.

dbfile$ **String.** The path and filename of the database within the Notes data directory. Use empty strings for both *dbfile$* and *server$* if you want to open the database later. Use a full path name if the database is not within the Notes data directory.

Return Value

notesDatabase   A NotesDatabase object that can be used to access the database you have specified.

If a database exists at the *server$* and *dbfile$* specified, the NotesDatabase object is open—and the script can access all its properties and methods. If a database does not exist at the *server$* and *dbfile$* specified, the NotesDatabase object is closed. To create a new database at the specified location, use this NotesDatabase object with the Create method.

Figure 4.1 shows a script example of creating a new database.

Figure 4.2 shows an example of creating a new database from a template.

Figure 4.3 shows an example of creating a replica copy.

Figure 4.4 shows an example of creating a non-replica database copy.

Figure 4.5 shows an example of deleting a database.

Figure 4.6 shows an example of opening a database.

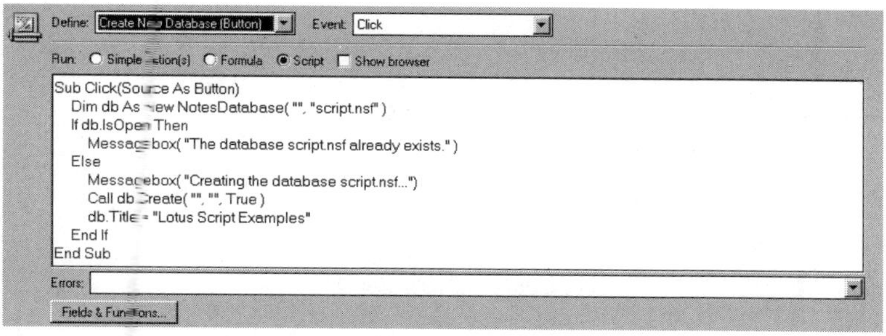
Define: Create New Database (Button)
Event: Click
Run: Simple action(s) Formula Script Show browser
Sub Click(Source As Button)
Dim db As New NotesDatabase("", "script.nsf")
If db.IsOpen Then
Messagebox("The database script.nsf already exists.")
Else
Messagebox("Creating the database script.nsf...")
Call db.Create("", "", True)
db.Title = "Lotus Script Examples"
End If
End Sub
Errors:
Fields & Functions...

Figure 4.1 Creating a new database script example

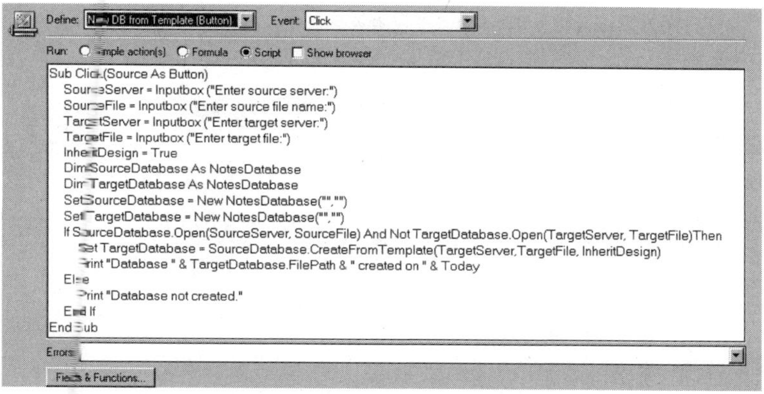
Define: New DB from Template (Button)
Event: Click
Run: Simple action(s) Formula Script Show browser
Sub Click(Source As Button)
SourceServer = Inputbox ("Enter source server:")
SourceFile = Inputbox ("Enter source file name:")
TargetServer = Inputbox ("Enter target server:")
TargetFile = Inputbox ("Enter target file:")
InheritDesign = True
Dim SourceDatabase As NotesDatabase
Dim TargetDatabase As NotesDatabase
Set SourceDatabase = New NotesDatabase("","")
Set TargetDatabase = New NotesDatabase("","")
If SourceDatabase.Open(SourceServer, SourceFile) And Not TargetDatabase.Open(TargetServer, TargetFile)Then
Set TargetDatabase = SourceDatabase.CreateFromTemplate(TargetServer,TargetFile, InheritDesign)
Print "Database " & TargetDatabase.FilePath & " created on " & Today
Else
Print "Database not created."
End If
End Sub
Errors:
Fields & Functions...

Figure 4.2 Creating a new database from a template

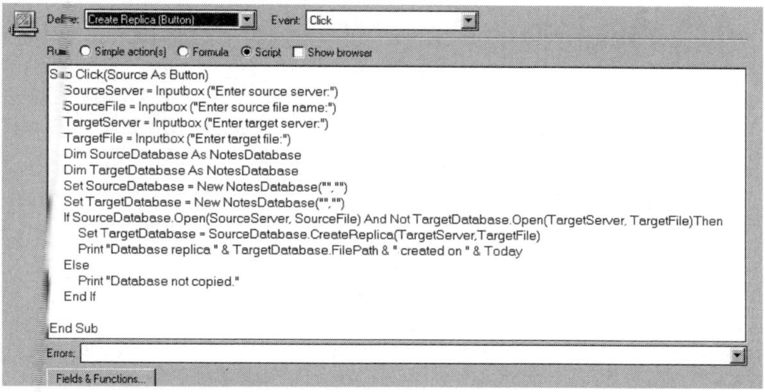
Define: Create Replica (Button)
Event: Click
Run: Simple action(s) Formula Script Show browser
Sub Click(Source As Button)
SourceServer = Inputbox ("Enter source server:")
SourceFile = Inputbox ("Enter source file name:")
TargetServer = Inputbox ("Enter target server:")
TargetFile = Inputbox ("Enter target file:")
Dim SourceDatabase As NotesDatabase
Dim TargetDatabase As NotesDatabase
Set SourceDatabase = New NotesDatabase("","")
Set TargetDatabase = New NotesDatabase("","")
If SourceDatabase.Open(SourceServer, SourceFile) And Not TargetDatabase.Open(TargetServer, TargetFile)Then
Set TargetDatabase = SourceDatabase.CreateReplica(TargetServer,TargetFile)
Print "Database replica " & TargetDatabase.FilePath & " created on " & Today
Else
Print "Database not copied."
End If
End Sub
Errors:
Fields & Functions...

Figure 4.3 Creating a replica copy

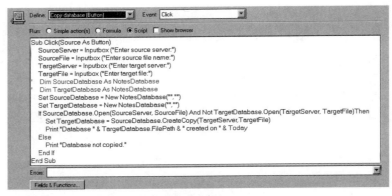

Figure 4.4 Creating a non-replica database copy

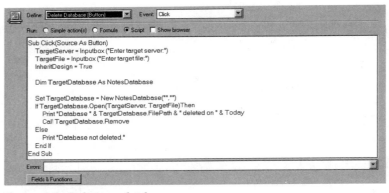

Figure 4.5 Deleting a database

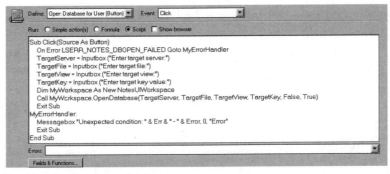

Figure 4.6 Opening a database

Note that the example in Figure 4.6 literally opens a database using the NotesWorkspace class OpenDatabase method. The difference between this and the NotesDatabase class Open method is that the Open method creates a new object instance of a database, which then enables you to access the database's properties and methods.

Note that the example in Figure 4.7 uses error-handling to catch any errors that occur.

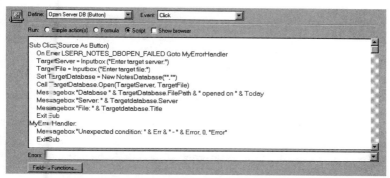

Figure 4.7   Error-handling

Database Information

Table 4.1 lists the NotesDatabase properties.

Table 4.1   NotesDatabase Properties

Property	Data Type	Description
ACL	NotesACL	Access control list for the database
Agents	NotesAgent array	Agents in the database
AllDocuments	NotesDocumentCollection	All documents in the database
Categories	String	(Read-write) Categories in the database

continues

Table 4.1 Continued

Property	Data Type	Description
Created	Date/Time	Date and time the database was created
CurrentAccessLevel	Integer	User's access level to the database
DelayUpdates	Boolean	(Read-write) True to delay (batch) server input/output for better performance
DesignTemplateName	String	Database's design template, if any
FileName	String	Database filename
FilePath	String	Database path
Forms	NotesForm array	Forms in the database
IsFTIndexed	Boolean	True if the database is full-text indexed
IsMultiDbSearch	Boolean	True if the search index is multi-database
IsOpen	Boolean	True if the database is open
IsPrivateAddressBook	Boolean	True if the database is a Personal Address Book; only valid through AddressBooks in NotesSession
IsPublicAddressBook	Boolean	True if the database is a Public Address Book; only valid through AddressBooks in NotesSession
LastFTIndexed	Date/Time	Date and time a full-text index, if any, was last updated

Property	Data Type	Description
LastModified	Date/Time	Date and time the database was last modified
Managers	String array	Users who have Manager access to the database
Parent	NotesSession	Current Notes session
PercentUsed	Double	Percentage of a database's total size that is occupied by real data
ReplicaID	String	Database replica ID in hexadecimal
Server	String	Name of the server containing the database
Size	Double	Database size, in bytes
SizeQuota	Long	(Read-write) Database size quota, if any; you must be an administrator to write
TemplateName	String	Database template name, if database is a template
Title	String	(Read-write) Database title
UnprocessedDocuments	NotesDocumentCollection	All documents not yet processed in agent or view
Views	NotesView array	Named views in the database

Sample Questions

1. *Objective:* Database access

   If you know a database's replica ID, you can open it using the

   a. OpenDatabase method
   b. New keyword
   c. OpenByReplicaID Method
   d. Create method

2. *Objective:* Database access

   A script needs at least _____ access to a database in order to open it.

   a. Depositor
   b. Reader
   c. Author
   d. Manager

3. *Objective:* Database creation

   To create a new database from scratch, use the _____ method.

   a. New keyword
   b. Open method
   c. Create
   d. None of the above

4. *Objective:* Database access

   A script that runs on a server

   a. Can access a database on another server
   b. Will not run
   c. Cannot open a database that is on a different server
   d. Will not access any database

5. *Objective:* Database access

The Open method

a. Opens a database in the user interface
b. Creates a new database
c. Creates a new instance of an NotesDatabase object and gives access to the database's properties and methods
d. Gives access to users of the script

6. *Objective:* Creating a new Notes database

The correct syntax for creating a new Notes database is

a. `Dim notesDatabase = New NotesDatabase( dbfile$, server$ )`
b. `Dim notesDatabase = New NotesDatabase( server$, replicaID$ )`
c. `Call notesDatabase.Create( server$, dbfile$, openFlag )`
d. `Dim notesDatabase = New NotesDatabase.Open( server$, dbfile$ )`

7. *Objective:* Creating a new replica of a Notes database

The correct syntax for creating a new Notes database is

a. `Set notesDatabase = notesDatabase.CreateReplica( newServer$, newDbFile$ )`
b. `Set notesDatabase = notesDatabase.CreateReplica( newServer$, replicaID$ )`
c. `Call notesDatabase.Replicate( server$, dbfile$)`
d. `Call notesDatabase.NewReplica( server$, dbfile$)`

8. *Objective:* Copying a Notes database

The correct syntax for copying a Notes database is

a. `Call notesDatabase.CreateCopy( newServer$, newDbFile$ )`
b. `Set notesDatabase = notesDatabase.CreateCopy( newDbFile$, newServer$ )`
c. `Call notesDatabase.Copy( server$, dbfile$)`
d. `Set notesDatabase = notesDatabase.CreateCopy( newServer$, newDbFile$ )`

9. *Objective:* Opening a Notes database object

The correct syntax for opening a Notes database object is

a. `flag = notesDatabase.Open( server$, dbfile$ )`

b. `Set notesDatabase = notesDatabase.Open( newDbFile$, newServer$ )`

c. `Call notesDatabase.OpenDb( server$, dbfile$)`

d. `flag = notesDatabase.Open(dbfile$, server$ )`

10. *Objective:* Opening a Notes database object

The correct syntax for opening a Notes database using a database's replica ID is

a. `Set notesDatabase = notesDatabase.OpenByReplicaID( server$, replicaID$ )`

b. `flag = notesDatabase.OpenByReplicaID( server$, replicaID$ )`

c. `flag = notesDatabase.OpenByReplicaID( server$, replicaID$ )`

d. `flag = notesDatabase.Open ( replica$, server$ )`

Sample Answers

1. *Answer:* c

   To access an existing database when you know its server and replica ID, use the OpenByReplicaID method.

2. *Answer:* b

   A script needs at least Reader access to a database in order to open it.

3. *Answer:* c

   To create a new database from scratch, use the Create method.

4. *Answer:* c

   A script that runs on a server cannot open a database that is on a different server. An error is returned if your script attempts to do so.

5. *Answer:* d

   The Open method creates a new object instance of a database. This command then enables you to access the database's properties and methods.

6. *Answer:* c

   The correct syntax for creating a new Notes database is `Call notesDatabase.Create( server$, dbfile$, openFlag )`. Using the New keyword only creates a new instance of a NotesDatabase object, not a physical database.

7. *Answer:* a

   The correct syntax for creating a new replica of a Notes database is `Set notesDatabase = notesDatabase.CreateReplica( newServer$, newDbFile$ )`. The Replicate method does not create a replica of a database, but rather it kicks off replication. There is no method called NewReplica.

8. *Answer:* d

   The correct syntax for creating a copy of a Notes database is `Set notesDatabase = notesDatabase.CreateCopy( newServer$, newDbFile$ )`.

9. *Answer:* a

   The correct syntax for creating a copy of a Notes database is `flag = notesDatabase.Open( server$, dbfile$ )`. You could also write `Call notesDatabase.Open( server$, dbfile$ )`.

10. *Answer:* d

    The correct syntax for opening a Notes database using a replica ID is `flag = notesDatabase.OpenByReplicaID( server$, replicaID$ )`.

CHAPTER 5

Finding Documents for Processing

This chapter reviews the various ways to find documents for processing and covers the NotesDocumentCollection class and accessing a document pool through the NotesView class. We also examine methods, such as the Search method and the FTSearch method, for retrieving document collections.

Chapter Objectives

The objectives of this chapter are to increase your understanding of

- Accessing all documents in a database
- Accessing documents through collections
- Searching for all documents in a database
- Searching for all documents in a view
- Processing a collection
- Full-text searching a collection
- Counting the documents in a collection

Document Pool by Collection

In LotusScript, there are a few ways to find documents for processing. If you know the UNID of a document, which is not usually known, you can find the document using the NotesDatabase class GetDocumentByUNID method. Or, if you do not know the document's identifier, you can iterate a collection or pool of documents to find the one you need.

The document pool can be obtained by a NotesCollection or NotesView and may consist of all or some of the documents in a database. Of the two classes, the NotesDocumentCollection offers more methods to develop the pool. Use a NotesDocumentCollection object if

- You want to act on a specific set of documents that meet certain criteria.
- There is no view in the database that contains every document you need to search.
- You do not need to navigate the documents' response hierarchies.

The NotesView class offers more flexibility in moving between documents—and includes methods that respect the hierarchical relationship between documents and response documents. Using views is a more efficient means of accessing documents, because they are already indexed by the database itself and take care of the selection and sorting ahead of time. They do not, however, necessarily provide access to the documents you want.

Use a NotesView object if

- There is a view or folder in the database that contains all the documents you want to search.
- You need to navigate through the documents' response hierarchies.
- You want to access documents as quickly as possible.
- You want to find a document by its key in a view.
- You want to access documents in sorted order.

NotesDocumentCollection Class

A NotesDocumentCollection represents a subset of all the documents in a database. The documents contained in the subset are determined by the NotesDatabase method or property that you use to search the database. The method or property can be any of the following:

- AllDocuments property
- UnprocessedDocuments property
- Search method
- UnprocessedSearch method
- FTSearch method
- UnprocessedFTSearch method

The NotesDocument class Responses property also returns a NotesDocumentCollection of the current document's responses. When using these methods to obtain a collection of documents, the documents in the collection are only ordered when they result from a full-text search. Otherwise, they are unordered.

The NotesDocumentCollection class has the following properties and methods for accessing documents:

- Count property and GetNthDocument method, which can be used to get documents in a For loop
- Get FirstDocument and GetNextDocument methods, which can be used to get the first and next documents in a While loop
- GetLastDocument and GetPrevDocument methods, which can be used to get the last and previous documents in a While loop

AllDocuments Property

The All Documents property of the NotesDatabase class returns a collection of documents equal to all of the documents in a database. The collection of all documents returned is an unsorted collection—and is a faster way of getting all the documents in the database, instead of using the NotesDatabase class Search method, where you would have to use the *Select @All* formula.

Syntax

```
Set notesDocumentCollection = notesDatabase.AllDocuments
```

AllDocuments Example

```
Sub Click(Source As Button)
'Declarations
    Dim Session As NotesSession
    Dim Db As NotesDatabase
    Dim Collection As NotesDocumentCollection
    Dim Doc As NotesDocument
    Dim j As Integer
'Set Object Variables
    Set Session = New NotesSession
    Set Db = Session.CurrentDatabase
'The document collection is gotten here using the AllDocuments
property
    Set Collection = Db.AllDocuments
'Loop through collection
    For j = 1 To Collection.Count
        Set Doc = Collection.GetNthDocument(j)
        Doc.Status = "Processed"
        Call Doc.Save(True,False)
    Next
End Sub
```

In this example, notice the use of the Count property of the NotesDocumentCollection class. This property returns the total number of documents contained in the collection. Because we know the number of documents in the collections, we can use a For/Next looping structure to loop through the documents—and we can use the NotesDocumentCollection class GetNthDocument method to get a handle on each Notes document as we loop through the collection.

UnprocessedDocuments Property

The NotesDatabase class UnprocessedDocuments property returns a collection of documents in a database that the current agent or view action considers to be *unprocessed*. The type of agent determines which documents are considered unprocessed. This method is valid only for agent or view action scripts and may be invoked only on NotesDatabase objects obtained from the CurrentDatabase property in NotesSession. In scripts that are not a part of an agent or view action, this method returns a NotesDocumentCollection with zero documents. When invoked on a NotesDatabase object that was not obtained from the CurrentDatabase property, this method returns an error.

Table 5.1 describes the documents that are returned by the UnprocessedDocuments property.

Table 5.1   Agent Settings and the UnprocessedDocuments Property

Agent Runs On	UnprocessedDocuments Returns Documents That Meet All of These Requirements
All documents in database	Meet the search criteria specified in Agent Builder
All new and modified documents since last run	Have not been processed by this agent with UpdateProcessedDoc Have just been created or modified Meet the search criteria specified in Agent Builder
All unread documents in view	Are unread and are in the view Meet the search criteria specified in Agent Builder
All documents in view	Are in the view Meet the search criteria specified in Agent Builder
Selected documents or View action	Are selected in the view Meet the search criteria specified in Agent Builder
Run once	Is the current document
Newly received mail documents	Have not been processed by this agent with UpdateProcessedDoc Have just been mailed to the database Meet the search criteria specified in Agent Builder

continues

Table 5.1   Continued

Agent Runs On	UnprocessedDocuments Returns Documents That Meet All of These Requirements
Newly modified documents	Have not been processed by this agent with UpdateProcessedDoc Have just been modified Meet the search criteria specified in Agent Builder
Pasted documents	Have not been processed by this agent with UpdateProcessedDoc Have just been pasted into the database Meet the search criteria specified in Agent Builder

For agents that run on new and modified documents, newly received mail documents, pasted documents, or newly modified documents, you must use the UpdateProcessedDoc method in NotesSession to mark each document as *processed*. This action assures that a document is processed by the agent only once (unless it is modified, mailed, or pasted again). If you do not call this method for each document, the agent processes the same documents the next time it runs.

- UpdateProcessedDoc marks a document as processed only for the particular agent from which it is called. Using UpdateProcessedDoc in one agent has no effect on the documents that another agent processes.

Syntax
```
Set notesDocumentCollection =
notesDatabase.UnprocessedDocuments
```

UnprocessedDocuments Examples
```
Sub Initialize
'Declarations
    Dim Session As NotesSession
    Dim Db As NotesDatabase
    Dim Collection As NotesDocumentCollection
```

```
            Dim Doc As NotesDocument
            Dim j As Integer
      'Set Object Variables
            Set Session = New NotesSession
            Set Db = Session.CurrentDatabase
      'The document collection is gotten here using the
      UnprocessedDocuments property
            Set Collection = Db.UnprocessedDocuments
      'Loop through collection
            For j = 1 To Collection.Count
                  Set Doc = Collection.GetNthDocument(j)
                  Doc.Status = "Open"
                  Call Doc.Save(True,False)
      'This method marks the document as processed so the agent will
      not select it the next time it runs
                  Call Session.UpdateProcessedDoc(Doc)
            Next
      End Sub
```

This example uses the UnprocessedDocuments property to get a document collection. The script uses the NotesSession class UpdateProcessedDoc method and could be used in an agent that runs on Newly Created or Modified documents.

Search Method

If the database that you are using to search for documents is not full-text indexed, you can use a selection formula (like a view selection formula) to refine the NotesDocumentCollection pool. The NotesDatabase class Search method returns all documents in a database that meet the criteria when given selection criteria for a document.

Syntax

```
      Set notesDocumentCollection = notesDatabase.Search( formula$,
      notesDateTime, maxDocs% )
```

Parameters

formula$ **String.**   A Notes @function formula that defines the selection criteria.

notesDateTime   A cutoff date. The method searches only documents created or modified since the cutoff date. This parameter is of the NotesDateTime class. In other words, you must create an object reference to a NotesDateTime class that holds the cutoff date you specify, and you should pass the object reference as a parameter to this method.

maxDocs% **Integer.**   The maximum number of documents you want returned. Specify zero to receive all matching documents.

Search Example

```
Sub Initialize
'Declarations
      Dim Session As NotesSession
      Dim Db As NotesDatabase
      Dim Collection As NotesDocumentCollection
      Dim Doc As NotesDocument
      Dim dateTime As NotesDateTime
      Dim j As Integer
'Set Object Variables
      Set Session = New NotesSession
      Set Db = Session.CurrentDatabase
      Set DateTime = New NotesDateTime("12/01/1996")
'Implicitly declare SearchFormula and assign it a search formula
      SearchFormula$ = "Select Form = ""Feedback"" & Status =
""Unanswered"""
'The document collection is gotten here using the Search method
      Set Collection = Db.Search(SearchFormula$, dateTime, 0)
'Loop through collection
      Set Doc = Collection.GetFirstDocument
      Do While Not (Doc Is Nothing)
            Doc.Status = "Answered"
            Call Doc.Save(True,False)
            Set Doc = Collection.GetNextDocument(Doc)
      Loop
End Sub
```

This example uses the NotesDatabase class Search method to retrieve a document collection. Notice the declaration of a NotesDataTime object. This declaration is necessary to use the Search method. The date that is specified should be the cutoff date for the documents you are trying to select. Also, the formula used in the Search method is the equivalent of a view selection formula. This method of selecting documents, when operating on a large set of documents, is considerably slower than using a NotesView object that references an existing view that selects the same documents. Technically, regardless of the amount of documents, this method will always be slower than accessing a NotesView object that references a view with a built index.

UnprocessedSearch

Given selection criteria for a document, the NotesDatabase class UnprocessedSearch method returns documents in a database which

- The current agent considers unprocessed
- Meet the search criteria
- Were created or modified since the cutoff date

This method works in two parts. First, it finds a collection of documents that the agent or view action considers *unprocessed*. The type of agent or view action determines which documents are considered unprocessed. This document collection is identical to that returned by the UnprocessedDocuments property.

Second, it conducts a search on the unprocessed documents and returns a collection of those documents that match the selection criteria and were created or modified since the cutoff date.

For example, in an agent that runs on all selected documents in the view, UnprocessedSearch searches only the selected documents and returns those that match the selection criteria. In an agent that runs on documents that have been created or modified since last run, UnprocessedSearch searches only the documents that have not been marked by the UpdateProcessedDoc method in NotesSession —and returns those that match the selection criteria.

This method is valid only for agent or view action scripts and may be invoked only on NotesDatabase objects obtained from the CurrentDatabase property in NotesSession. In scripts that are not part of an agent or view action, this method returns a NotesDocumentCollection with zero documents. When invoked on a NotesDatabase object that was not obtained from the CurrentDatabase property, this method returns an error.

Syntax

```
Set notesDocumentCollection = notesDatabase.UnprocessedSearch(
formula$, notesDateTime, maxDocs% )
```

Parameters

formula$ **String.** A Notes @function formula that defines the selection criteria.

notesDateTime A cutoff date. The method searches only documents created or modified since the cutoff date.

maxDocs% **Integer.** The maximum number of documents you want returned. Set this parameter to zero to receive all matching documents.

Return Value

```
notesDocumentCollection
```

A collection of documents that are unprocessed, match the selection criteria, and are created or modified after the cutoff date. The collection is sorted by relevance, with highest relevance first.

Table 5.2 describes the documents that are returned by the UnprocessedSearch method.

Table 5.2 Agent Settings and UnprocessedSearch Method

Agent Runs On	UnprocessedSearch Returns Documents That Meet All of These Requirements
All documents in database	Meet the search criteria specified in Agent Builder Meet the @function search critieria specified in this method
All new and modified documents since last run	Have not been processed by this agent with UpdateProcessedDoc Have just been created or modified Meet the search criteria specified in Agent Builder Meet the @function search critieria specified in this method
All unread documents in view	Are unread and in the view Meet the search criteria specified in Agent Builder Meet the @function search criteria specified in this method
All documents in view	Are in the view Meet the search criteria specified in Agent Builder Meet the @function search criteria specified in this method
Selected documents or View action	Are selected in the view Meet the search criteria specified in Agent Builder Meet the @function search criteria specified in this method
Run once	Is the current document

Agent Runs On	UnprocessedSearch Returns Documents That Meet All of These Requirements
Newly received mail documents	Have not been processed by this agent with UpdateProcessedDoc Have just been mailed to the database Meet the search criteria specified in Agent Builder Meet the @function search criteria specified in this method
Newly modified documents	Have not been processed by this agent with UpdateProcessedDoc Have just been modified Meet the search criteria specified in Agent Builder Meet the @function search criteria specified in this method
Pasted documents	Have not been processed by this agent with UpdateProcessedDoc Have just been pasted into the database Meet the search criteria specified in Agent Builder Meet the @function search criteria specified in this method

For agents that run on new and modified documents, newly received mail documents, pasted documents, or newly modified documents, you must use the UpdateProcessedDoc method in the NotesSession class to mark each document as *processed*. This action assures that a document gets processed by the agent only once (unless it is modified, mailed, or pasted again). If you do not call this method for each document, the agent processes the same documents the next time it runs.

- UpdateProcessedDoc marks a document as processed only for the particular agent from which it is called. Using UpdateProcessedDoc in one agent has no effect on the documents that another agent processes.

- In all other agents and view actions, UpdateProcessedDoc has no effect.

UnprocessedSearch Example

```
Sub Initialize
'Declarations
      Dim Session As NotesSession
      Dim Db As NotesDatabase
      Dim Collection As NotesDocumentCollection
      Dim Doc As NotesDocument
      Dim dateTime As NotesDateTime
      Dim j As Integer
'Set Object Variables
      Set Session = New NotesSession
      Set Db = Session.CurrentDatabase
      Set DateTime = New NotesDateTime("12/01/1996")
'Implicitly declare SearchFormula and assign it a search formula
      SearchFormula$ = "Select Form = ""Feedback"" &
@Contains(""Subject"";""Stocks and Bonds"")"
'The document collection is gotten here using the      .
UnprocessedSearch method
      Set Collection = Db.UnprocessedSearch(SearchFormula$,
dateTime, 0)
'Loop through collection
      Set Doc = Collection.GetFirstDocument
      Do While Not (Doc Is Nothing)
            Call Doc.PutInFolder("Market Analysis")
            Set Doc = Collection.GetNextDocument(Doc)
      Loop
End Sub
```

This agent script runs on all unread documents created after Dec. 1, 1996, and contain *Stocks and Bonds* in the Subject item. The script's purpose is to put these documents into a folder.

- If the agent does not contain a search defined in the Agent Builder, UnprocessedSearch returns all unread documents in the database that were created after Dec. 1, 1996, and contain *Stocks and Bonds* in the Subject item, regardless of whether the agent has already run on some of the unread documents at an earlier time.

- If the agent does contain searches defined in the Agent Builder, UnprocessedSearch returns all unread documents in the database that meet the search criteria defined in the Agent Builder, that were created after Dec. 1, 1996, and that contain the *Stocks and*

Bonds, regardless of whether the agent has already run on some of the unread documents at an earlier time.

FTSearch Method

The FTSearch method conducts a full-text search. This method is contained in both the NotesDatabase, NotesDocumentCollection and the NotesView classes. In the NotesDatabase class, the search will return a subset of all of the documents in the database that match the search criteria. In the NotesDocumentCollection class, the FTSearch method conducts a full-text search of all the documents in a Notes database collection and reduces the collection to those documents that match. In the NotesView class, the FTSearch method conducts a full-text search against the documents in the referenced view and returns a collection of documents that match the search criteria.

Syntax

```
NotesDatabase Class
Set notesDocumentCollection = notesDatabase.FTSearch( query$,
maxDocs% [,sortoptions [, otheroptions]] )
NotesDocumentCollection Class
Call notesDocumentCollection =
notesDocumentCollection.FTSearch( query$, maxDocs%)
NotesView Class
numDocs% = notesView.FTSearch( query$, maxDocs% )
```

Parameters

query$ **String.** The full-text query. Read on for the syntax.

maxDocs% **Integer.** The maximum number of documents you want returned from the query. Set this parameter to zero to receive all matching documents.

sortoptions **Integer.** Optional. Use one of three constants to specify a sorting option

- FT_SCORES (default) sorts by relevance score.
- FT_DATE_DES sorts by document creation date in descending order.
- FT_DATE_ASC sorts by document creation date in ascending order.

otheroptions **Integer.** Optional. Use one of two constants to specify additional search options

- FT_STEMS uses stem words as the basis of the search.
- FT_THESAURUS uses the thesaurus to search.

You can use both options together by specifying FT_STEMS + FT_THESAURUS

Return Value
```
notesDocumentCollection
```

A collection of documents that match the full-text query, sorted by the selected option. When the collection is sorted by relevance, the highest relevance appears first. You can access the relevance score of each document in the collection using the FTSearchScore property in NotesDocument.

If the database is not full-text indexed, this method works—but less efficiently. To test for an index, use the IsFTIndexed property. To create an index on a local database, use the UpdateFTIndex method.

This method searches all of the documents in a database. To search only documents found in a particular view, use the FTSearch method in NotesView. To search only documents found in a particular document collection, use the FTSearch method in NotesDocumentCollection.

If you do not specify any sort options, you receive the documents sorted by relevance score. If you ask for a sort by date, you do not get relevance scores. If you pass the resulting DocumentCollection to a NotesNewsletter instance, it formats its doclink report with either the document creation date or the relevance score, depending on which sort options you use.

If the database has a multi-database index, you get a multi-database search. Although navigating through the resulting document collection can be slow, you can efficiently create a newsletter from the collection.

Full-Text Query Syntax
The syntax rules for a search query are as follows. (Use parentheses to override precedence and to group operations.)

- Plain text—To search for a word or phrase, enter the word or phrase as is—but search keywords and symbols must be enclosed in quotes. To be on the safe side, enclose all search text in quotes. Remember to use double quotes if you are inside a LotusScript literal.

- Wildcards—Use a question mark (?) to match any single character in any position in a word. Use an asterisk (*) to match zero-to-many characters in any position in a word.

- Hyphenated words—Use hyphenated words to find two-word pairs that are hyphenated, that run together as a single word, or that are separated with a space.

- Logical operators—Use logical operators to expand or restrict your search. The operators and their precedents are not (!), and (&), accrue (,), and and/or (|). You can use either the keyword or symbol.

- Proximity operators—Use proximity operators to search for words that are close to each other. These operators require word, sentence, and paragraph breaks in a full-text index. The operators are near, sentence, and paragraph.

- Field operator—Use the field operator to restrict your search to a specified field. The syntax is FIELD *field-name operator*, where *operator* contains text and rich text fields and is one of the following for number and date fields: =, >, >=, <, or <=.

- Exactcase operator—Use the exactcase operator to restrict a search for the next expression to the specified case.

- Termweight operator—Use the termweight *n* operator to adjust the relevance ranking of the expression that follows, where *n* is 0–100.

FTSearch Examples
NotesDatabase Class

```
Sub Initialize
'Declarations
      Dim Session As New NotesSession
      Dim Db As NotesDatabase
      Dim Collection As NotesDocumentCollection
      Dim Newsletter As NotesNewsletter
      Dim Doc As NotesDocument
'Set Objects
      Set Db = Session.CurrentDatabase
'Full Text search the current database for any document
containing the phrase "physical therapy"
      If Db.IsFTIndexed Then
            Set Collection = Db.FTSearch("""physical therapy""",
0, FT_SCORES, FT_STEMS)
            Set Newsletter = New NotesNewsletter( collection )
            Set Doc = newsletter.FormatMsgWithDoclinks( Db )
            Doc.Form = "Memo"
            Doc.Subject = "Here's the newsletter you requested."
            Call Doc.Send( False, """Jennifer Hoos""" )
      End If
End Sub
```

This script searches the current database for the phrase *physical therapy*. Every document containing the phrase is placed into the collection in order of relevance (FT_SCORES), and then is formatted into a newsletter.

NotesDocumentCollection Class

```
Sub Initialize
'Declarations
     Dim Session As New NotesSession
     Dim Db As NotesDatabase
     Dim Collection As NotesDocumentCollection
     Dim Newsletter As NotesNewsletter
     Dim Doc As NotesDocument
     Dim dateTime As NotesDateTime
'Set Objects
     Set Db = Session.CurrentDatabase
     Set dateTime = New NotesDateTime("12/01/97")
     SearchFormula$ = "Select Form = ""MedicalFacts"""
     Set Collection = Db.Search(SearchFormula$,notesDateTime,0)
'Full Text search the document collection for any document
containing the phrase "physical therapy"
     Set FTCollection = Collection.FTSearch("""physical
therapy""", 0)
     Set Newsletter = New NotesNewsletter( FTCollection )
     Set Doc = newsletter.FormatMsgWithDoclinks( Db )
     Doc.Form = "Memo"
     Doc.Subject = "Here's the newsletter you requested."
     Call Doc.Send( False, """Maria Teresa Rodriguez""" )
End Sub
```

This script first gets a document collection by using the NotesDatabase search method. The script then performs a full-text search on the collection to return a subset of documents, which is formatted into a newsletter and is mailed.

NotesView Class

```
Sub Initialize
'Declarations
     Dim Db As NotesDatabase
     Dim View As NotesView
     Dim j As Integer
'Set objects
     Set Db = New NotesDatabase( "TestServer01",
"ScriptTalk.nsf" )
     Set view = db.GetView( "By Date\Ascending By Main Topic" )
     j = view.FTSearch( "benchmark*", 0 )
End Sub
```

In this example, the FTSearch method returns the number of documents that meet the specified criteria in the view.

UnprocessedFTSearch Method

Given selection criteria for a document, this method returns documents in a database which

- The current agent considers to be unprocessed
- Match the query

Just like the UnprocessedDocument method, this method works in two parts. First, it finds a collection of documents that the agent considers to be *unprocessed*. The type of agent or view action determines which documents are considered to be unprocessed. This document collection is identical to that returned by the UnprocessedDocuments property.

Second, it conducts a full-text search on the unprocessed documents and returns a collection of those documents which match the query.

For example, in an agent that runs on all selected documents in the view, UnprocessedFTSearch searches only the selected documents and returns those that match the query. In an agent that runs on documents that have been created or modified since last run, UnprocessedFTSearch searches only the documents that have not been marked by the UpdateProcessedDoc method in NotesSession— and returns those which match the query.

Syntax

```
Set notesDocumentCollection = notesDatabase.UnprocessedFTSearch(
query$, maxDocs% [,sortoptions [, otheroptions]] )
```

Parameters

query$ **String.**   The full-text query.

maxDocs% **Integer.**   The maximum number of documents you want returned from the query. Set this parameter to zero to receive all matching documents.

sortoptions **Integer.**   Optional. Use one of three constants to specify a sorting option

- FT_SCORES (default) sorts by relevance score.
- FT_DATE_DES sorts by document creation date in descending order.
- FT_DATE_ASC sorts by document creation date in ascending order.

otheroptions **Integer.** Optional. Use one of two constants to spec-ify additional search options:

- FT_STEMS uses stem words as the basis of the search.
- FT_THESAURUS uses the thesaurus to search.

You can use both options together by specifying FT_STEMS + FT_THESAURUS.

This method is valid only for agent scripts and view actions—and may be invoked only on NotesDatabase objects obtained from the CurrentDatabase property in NotesSession. In scripts that are not part of an agent or view action, this method returns a NotesDocumentCollection with zero documents. When invoked on a NotesDatabase object that was not obtained from the CurrentDatabase property, this method raises an error. If the data-base is not full-text indexed, this method works—but less effi-ciently. To test for an index, use the IsFTIndexed property. To create an index on a local database, use the UpdateFTIndex method.

Table 5.3 describes the documents that are returned by the UnprocessedFTSearch method.

Table 5.3 Agent Settings and the UnprocessedFTSearch Method

Agent Runs On	UnprocessedSearch Returns Documents That Meet All of These Requirements
All documents in database	Meet the search criteria specified in Agent Builder Meet the full-text search critieria specified in this method
All new and modified documents since last run	Have not been processed by this agent with UpdateProcessedDoc Have just been created or modified Meet the search criteria specified in Agent Builder Meet the full-text search critieria specified in this method

Agent Runs On	UnprocessedSearch Returns Documents That Meet All of These Requirements
All unread documents in view	Are unread and are in the view Meet the search criteria specified in Agent Builder Meet the full-text search criteria specified in this method
All documents in view	Are in the view Meet the search criteria specified in Agent Builder Meet the full-text search criteria specified in this method
Selected documents or View action	Are selected in the view Meet the search criteria specified in Agent Builder Meet the full-text search criteria specified in this method
Run once	Is the current document
Newly received mail documents	Have not been processed by this agent with UpdateProcessedDoc Have just been mailed to the database Meet the search criteria specified in Agent Builder Meet the full-text search criteria specified in this method
Newly modified documents	Have not been processed by this agent with UpdateProcessedDoc Have just been modified Meet the search criteria specified in Agent Builder Meet the full-text search criteria specified in this method

continues

Table 5.3   Continued

Agent Runs On	UnprocessedSearch Returns Documents That Meet All of These Requirements
Pasted documents	Have not been processed by this agent with UpdateProcessedDoc Have just been pasted into the database Meet the search criteria specified in Agent Builder Meet the full-text search criteria specified in this method

Using UpdateProcessedDoc

For agents that run on new and modified documents, newly received mail documents, pasted documents, or newly modified documents, you must use the UpdateProcessedDoc method in NotesSession to mark each document as *processed*, which assures that a document gets processed by the agent only once (unless it is modified, mailed, or pasted again). If you do not call this method for each document, the agent processes the same documents the next time it runs.

■ UpdateProcessedDoc marks a document as processed only for the particular agent from which it is called. Using UpdateProcessedDoc in one agent has no effect on the documents that another agent processes.

In all other agents and view actions, UpdateProcessedDoc has no effect.

UnprocessedFTSearch Example

```
Sub Initialize
'Declarations
     Dim Session As New NotesSession
     Dim Db As NotesDatabase
     Dim Collection As NotesDocumentCollection
     Dim Newsletter As NotesNewsletter
     Dim Doc As NotesDocument
'Set Objects
     Set Db = Session.CurrentDatabase
'Full Text search the current database for any document
```

```
containing the phrase "physical therapy"
     If Db.IsFTIndexed Then
          Set Collection = Db.UnprocessedFTSearch("""physical
therapy""", 0, FT_SCORES, FT_STEMS)
          Set Newsletter = New NotesNewsletter( collection )
          Set Doc = newsletter.FormatMsgWithDoclinks( Db )
          Doc.Form = "Memo"
          Doc.Subject = "Here's the newsletter you requested."
          Call Doc.Send( False, """John Pionke""" )
     End If
End Sub
```

This script searches the current database for unprocessed documents that contain the phrase *physical therapy*. Every document containing the phrase is placed into the collection in order of relevance (FT_SCORES), and then is formatted into a newsletter.

Document Pool by View

Instead of selecting from all documents in a database, you can access a pool of documents using a view or a folder. The view is usually a subset of the documents in the database. You can

- Rely on its selection formula and sorted columns
- Distinguish between the three types of documents (document, response-to-document, response-to-response) and navigate appropriately

Document/Document Pool by Key

You can further refine the selection of the first document from a view-generated pool by using a key. When using a key, you can either search for a specific document or return a collection of documents. The two methods in the NotesView class are

- GetDocumentByKey
- GetAllDocumentsByKey

GetDocumentByKey Method

The GetDocumentByKey method finds a document within a view based on its column values. You create an array of strings (keys), where each key corresponds with a value in a sorted column in the view. The method returns the first document whose column values match each key in the array. To locate all matching documents, use GetAllDocumentsByKey—or use this method followed by a GetNextDocument loop.

Syntax

```
Set notesDocument = notesView.GetDocumentByKey( keyArray [ ,
exact ] )
```

Parameters

keyArray **String or Array of Strings.** Each element in the array contains a string, which is compared to a sorted column in the view. The first element in the array is compared with the first sorted column in the view; the second element is compared to the second sorted column; and so on.

exact **Boolean.** Optional. Specify True if you want to find an exact match. If you specify False (the default) or omit this parameter, a partial match succeeds.

GetDocumentByKey Example

```
Sub Initialize
'Declarations
      Dim Session As NotesSession
      Dim Db As NotesDatabase
      Dim View As NotesView
      Dim Doc As NotesDocument
      Dim PersonName As String
'Set objects
      Set Session = New NotesSession
      Set Db = Session.CurrentDatabase
      Set view = Db.GetView( "HRPersonnel" )
'Promp user for a name
      PersonName = Inputbox$("Enter an employee's
name","Employee Name", Session.CommonUserName)
'Find document based on the key PersonName
      Set Doc = View.GetDocumentByKey(PersonName)
      If Not (Doc Is Nothing) Then
            Messagebox PersonName & " has a salary of:  " &
Doc.Salary(0), 48, "Salary"
      Else
            Messagebox "The person you entered does not exist!",
16, "Error!"
      End If
End Sub
```

In this example, we prompt the user for an employee's name. We then use that name as a key to look up a document in the *HREmployees* view. If there are multiple employees with the same name, this will only return the first document that matches the key in the view.

GetAllDocumentsByKey Method

The GetAllDocumentsByKey method finds documents based on their column values within a view. You create an array of strings (keys), where each key corresponds with a value in a sorted column in the view. This method returns all documents whose column values match each key in the array. If no documents match, the collection is empty—and the collection count is zero.

Syntax

```
Set notesDocumentCollection = notesView.GetAllDocumentsByKey(
keyArray [ ,exact ] )
```

Parameters

keys **String or array of strings.** Each element in the array contains a string, which is compared with a sorted column in the view. The first element in the array is compared with the first sorted column in the view; the second element is compared with the second sorted column; and so on.

exact **Boolean.** Optional. Specify True if you want to find an exact match. If you specify False (the default) or omit this parameter, a partial match succeeds.

GetAllDocumentsByKey Example

```
Sub Initialize
'Declarations
    Dim Session As NotesSession
    Dim Db As NotesDatabase
    Dim View As NotesView
    Dim Collection As NotesDocumentCollection
    Dim Doc As NotesDocument
    Dim Department As String
'Set objects
    Set Session = New NotesSession
    Set Db = Session.CurrentDatabase
    Set view = Db.GetView( "CompanyDepartments" )
'Prompt user for a name
    Department = Inputbox$("Enter a department","Department
Salary")
'Find all employee documents based on the department key in the
CompanyDepartments view
    Set Collection = View.GetAllDocumentsByKey(Department)
    For j = 1 To Collection.Count
        Set Doc = Collection.GetNthDocument(j)
        Messagebox Doc.PersonName(0) & " has a salary of:  "
& Doc.Salary(0), 48, "Salary"
    Next
End Sub
```

In this example, we prompt the user for a department name. Based on that department name, we use the GetAllDocumentsByKey method to return all documents from the *CompanyDepartments* view that match that key. The script then loops through the collection and displays a message box listing each employee's salary.

Sample Questions

1. *Objective:* NotesDocumentCollection class

   You should use a NotesDocumentCollection when searching for a pool of documents if

   a. You want to act on a specific set of documents that meet certain criteria.

   b. There is no view in the database that contains every document you need to search.

   c. You do not need to navigate the documents' response hierarchies.

   d. All of the above

2. *Objective:* NotesView class

   You should use a NotesView object if

   a. You do not need to navigate through a document's response hierarchy.

   b. Quickness is not an issue.

   c. You want to access documents in sorted order.

   d. The documents you are trying to pool are not in a view or folder.

3. *Objective:* NotesDocumentCollection class

   A method or property that returns a document collection is

   a. A NotesDatabase class Search method

   b. A NotesDatabase FTSearch method

   c. A NotesSesson AllDocuments property

   d. All of the above

4. *Objective:* NotesDocumentCollection class

   The NotesDocumentCollection class has the following properties and methods for accessing documents:

   a. GetAllDocuments
   b. GotoNextDocument
   c. GetNthDocument
   d. FindNextDocument

5. *Objective:* NotesDatabase class Search Method

   The Search method

   a. Only works on a full-text indexed database
   b. Uses a key as a parameter to select documents
   c. Uses a Notes @function formula that defines the selection criteria
   d. Only returns the first document found

6. *Objective:* UnprocessedSearch method

   Given selection criteria for a document, the NotesDatabase class UnprocessedSearch method returns documents in a database which

   a. The current agent considers unprocessed
   b. Meet the search criteria
   c. Were created or modified since the cutoff date
   d. All of the above

7. *Objective:* Document/document pool by key

   The methods GetDocumentByKey and GetAllDocumentsByKey belong to which class?

   a. NotesSession
   b. NotesDatabase
   c. NotesDocumentCollection
   d. None of the above

8. *Objective:* Document/document pool by key

   True or False: The method GetDocumentByKey returns a collection of documents just like @DbColumn does.

9. *Objective:* Document pool by view

   An advantage of using a NotesView to get a document pool is

   a. You can rely on its selection formula and sorted columns.

   b. A NotesView is technically faster than accessing documents through a NotesDocumentCollection.

   c. Being able to distinguish between the three types of documents (document, response-to-document, response-to-response) and navigate appropriately

   d. All of the above

10. *Objective:* NotesView GetAllDocumentsByKey method

    If no documents match the key parameter specified for the GetAllDocumentsByKey method

    a. The collection is empty, and the collection count is zero.

    b. An error occurs, and you must use error-handling to catch it.

    c. The collection is empty, and the collection count is one.

    d. The collection is empty, and the collection count property is unavailable.

11. *Objective:* Count a collection

    To get the total number of documents in a DocumentCollection, use the

    a. Count property

    b. NumberOfDocs property

    c. AllDocuments property

    d. Loop through the collection and count the documents one-by-one

Sample Answers

1. *Answer:* d

   Use a NotesDocumentCollection object if

   - You want to act on a specific set of documents that meet certain criteria.
   - There is no view in the database that contains every document you need to search.
   - You do not need to navigate the documents' response hierarchies.

2. *Answer:* c

   Use a NotesView object if

   - There is a view or folder in the database that contains all the documents you want to search.
   - You need to navigate through the documents' response hierarchies.
   - You want to access documents as quickly as possible.
   - You want to find a document by its key in a view.
   - You want to access documents in sorted order.

3. *Answer:* d

4. *Answer:* c

   The NotesDocumentCollection class has the following properties and methods for accessing documents

   - Count property and GetNthDocument method, which can be used to get documents in a For loop
   - Get FirstDocument and GetNextDocument methods, which can be used to get the first and next documents in a While loop
   - GetLastDocument and GetPrevDocument methods, which can be used to get the last and previous documents in a While loop

5. *Answer:* c

   Given selection criteria for a document, the NotesDatabase class Search method returns all documents in a database that meet the criteria. The formula parameter uses a Notes @function formula that defines the selection criteria.

6. *Answer:* d

   Given selection criteria for a document, the NotesDatabase class UnprocessedSearch method returns documents in a database which

   - The current agent considers unprocessed
   - Meet the search criteria
   - Were created or modified since the cutoff date

7. *Answer:* d

   The methods GetDocumentByKey and GetAllDocumentsByKey belong to the NotesView class.

8. *Answer:* False

   The NotesView method GetDocumentByKey returns only the first document that matches the specified key parameter — even if there are multiple documents in the view with the same key.

9. *Answer:* d

   Instead of selecting from all documents in a database, you can access a pool of documents using a view or a folder. The view is usually a subset of the documents in the database, and you can

   - Rely on its selection formula and sorted columns
   - Distinguish between the three types of documents (document, response-to-document, response-to-response) and navigate appropriately

   NotesViews are technically much faster because they already contain an index of documents — which a NotesDocumentCollection does not.

10. *Answer:* a

    If no documents match, the collection is empty — and the collection count is zero.

11. *Answer:* a

    To retrieve the total number of documents in a DocumentCollection, use the Count property.

CHAPTER 6

Processing Documents

In the previous chapter, we reviewed how to identify documents for processing. This chapter reviews the different ways you can process those documents. Included in this chapter are how to create, copy, and delete documents, as well as send mail and create newsletters. We will review the NotesDocument class, some of its properties and methods, and the NotesNewsletter class.

Chapter Objectives

The objectives of this chapter are to increase your understanding of the following items:

- Creating documents
- Copying documents
- Deleting documents
- Making response documents
- Putting documents in folders
- Sending mail
- Creating newsletters

NotesDocument Class

The NotesDocument class lets you examine and manipulate document properties and contents. You gain access to a NotesDocument object through methods in the NotesDatabase, NotesView, and NotesDocumentCollection classes.

Creating New Documents

Use the New method of a NotesDocument class or the Create-Document method from the NotesDatabase class to create a new document in a database. Both methods will create a new instance of a NotesDocument object reference, but this does not mean the document has been saved to the database. This fact simply means that there is a placeholder in memory for it. It is not until you use the Save method of the NotesDocument class that it is saved to disk. If you do not use the Save method, the new document is lost when the program exits. There are, in essence, four steps to creating a Notes document using LotusScript:

1. Declaring your New NotesDocument object
2. Adding items (fields)
3. Setting the form name (which is just another item)
4. Calling the NotesDocument object Save method, which has two parameters to prevent overwriting an existing document. In the case of a new document, the parameters should be set to True, False. For example: Call Doc.Save(True, False).

You can also create documents using the user interface. The NotesUIWorkspace has the ComposeDocument method, which functions the same way @Command([Compose]) does. Using the form you specify, it will display it to the user for data entry.

NotesDocument Class New Method

Given a database, New creates a document in the database and returns a NotesDocument object that represents the document.

Syntax

```
Dim variableName As New NotesDocument( notesDatabase )
```

or

```
Set notesDocument = New NotesDocument( notesDatabase )
```

Parameter

`notesDatabase` The database in which to create a new document.

NotesDatabase Class CreateDocument Method

```
Set notesDocument = notesDatabase.CreateDocument
```

Return Value

`notesDocument` The newly-created document.

NotesUIWorkspace Class ComposeDocument Method

```
Set notesUIDocument = notesUIWorkspace.ComposeDocument( [
server$ [ , file$ [ , form$ [ , windowWidth# [ , windowHeight#
]]] ] ] )
```

Parameters

`server$` **String.** Optional. The name of the server where the database resides. If this parameter is omitted or is an empty string (" "), the database is opened on the local computer. If both server$ and file$ are omitted or empty strings, the current database is used.

`file$` **String.** Optional. The file name of the database you want to compose a document in. If this parameter is omitted or is an empty string (" "), a document is composed in the current database.

`form$` **String.** Optional. The name of the form you want to use to compose the document. If this parameter is omitted or is an empty string (" "), the Create Other dialog box is displayed and the user can select a form name.

`windowWidth#` **Double.** Optional. The width in inches of the document window. If this parameter is omitted, the window is the width of the current window.

`windowHeight#` **Double.** Optional. The height in inches of the document window. If this parameter is omitted, the window is the height of the current window.

Return Value
`notesUIDocument`   The document that was just created and opened.

Creating New Documents Examples

NotesDocument New Example

This example creates a document in the current database.

```
Sub Click(Source As Button)
     On Error Goto ErrorHandler
'Declarations
     Dim Session As NotesSession
     Dim Db As NotesDatabase
     Dim newDoc As NotesDocument
'Object creation
     Set Session = New NotesSession
     Set Db = Session.CurrentDatabase
'The NotesDocument object reference is set in memory here using
the New method
     Set newDoc = New NotesDocument(Db)
'Check to see if a new object reference was created
     If Not (newDoc Is Nothing) Then
'Add some NotesItems using the extended syntax
          newDoc.Name = "Bill"
          newDoc.Address = "1234 N. South Street"
          newDoc.City = "Chicago"
          newDoc.State = "IL"
'Set the form name with this statement
          newDoc.Form = "Profile"
'Save the document
          Call newDoc.Save(True,False)
     Else
          Messagebox "There were problems wit the creation
process."
     End If
     Exit Sub
ErrorHandler:
     Messagebox "There were problems with the creation
process."
End Sub
```

Note that in this example, there is an example of error-handling and data validation. The If . . . Then statement checks to see if the newDoc object has been set. If it has not, we can assume that there was an error with the creation of the document. The On Error statement will trap any other errors that occur.

Creating a Document in Another Database

This example creates a document in another database:

```
Sub Click(Source As Button)
    On Error Goto ErrorHandler
'Declarations
    Dim Db As NotesDatabase
    Dim newDoc As NotesDocument
'Object creation
    Set Db = New NotesDatabase("","")
    openFlag = Db.OpenByReplicaID( "", "85255FA900747B84" )
'Check to see if the database was opened
    If openFlag = True Then
'The NotesDocument object reference is set in memory here using
the New method
        Set newDoc = New NotesDocument(Db)
'Check to see if a new object reference was created
        If Not (newDoc Is Nothing) Then
'Add some NotesItems
            newDoc.Name = "Bill"
            newDoc.Address = "1234 N. South Street"
            newDoc.City = "Chicago"
            newDoc.State = "IL"
'Set the form name with this statement
            newDoc.Form = "Profile"
'Save the document
            Call newDoc.Save(True,False)
        Else
            Messagebox "There was an error creating the
document."
        End If
    Else
        Messagebox "The database could not be open."
    End If
    Exit Sub
ErrorHandler:
    Messagebox "There were problems with the creation
process."
End Sub
```

In this example, we use the NotesDatabase OpenByReplicaID method to open another database. Notice that the replica ID specified omits the colon from it. We then add another line of data validation to check to see whether the database was opened. The only difference from Example 1 and Example 2 is that the Db object reference points to different databases. The creation of the Notes document is the same.

NotesDatabase CreateDocument Example

```
Sub Click(Source As Button)
    On Error Goto ErrorHandler
'Declarations
    Dim Db As NotesDatabase
    Dim newDoc As NotesDocument
'Object creation
    Set Db = New NotesDatabase("","")
    openFlag = Db.OpenByReplicaID( "", "85255FA900747B84" )
'Check to see if the database was opened
    If openFlag = True Then
'The NotesDocument object reference is set in memory here using
the CreateDocument Method
        Set newDoc = Db.CreateDocument
'Check to see if a new object reference was created
        If Not (newDoc Is Nothing) Then
'Add some NotesItems
            newDoc.Name = "Bill"
            newDoc.Address = "1234 N. South Street"
            newDoc.City = "Chicago"
            newDoc.State = "IL"
'Set the form name with this statement
            newDoc.Form = "Profile"
'Save the document
            Call newDoc.Save(True,False)
        Else
            Messagebox "There was an error creating the
document."
        End If
    Else
        Messagebox "The database could not be open."
    End If
    Exit Sub
ErrorHandler:
    Messagebox "There were problems with the creation process."
End Sub
```

NotesUIWorkspace Class ComposeDocument Example

```
Sub Click(Source As Button)
'Declare and set new NotesUIWorkspace object
    Dim workspace As New NotesUIWorkspace
'Compose a new document
    Call workspace.ComposeDocument( "", "", "Main Topic" )
End Sub
```

Creating Response Documents

To create response documents in LotusScript, the parent and child documents (or what will become the child) must be in the same database. A lot of times, this method is used to create relationships between existing documents (as opposed to making a document a response to another document, while using the NotesDocument class New method). The most challenging part of doing this is identifying the documents for processing. It is best to use clear variable naming practices to make your code more legible and easier to understand for yourself and for other developers. Just as creating a new document, you must call the NotesDocument class Save method when you make a document a response.

Syntax
```
Call notesDocument.MakeResponse( parentDocument )
```

Parameter

parentDocument   The document to which the current document becomes a response.

Creating Response Documents Examples

Making the Current Document a Response Document
```
Sub Click(Source As Button)
    On Error Goto ErrorHandler
'Declarations
    Dim Workspace As NotesUIWorkspace
    Dim UIDoc As NotesUIDocument
    Dim Session As NotesSession
    Dim Db As NotesDatabase
    Dim View As NotesView
    Dim parentDoc As NotesDocument
    Dim Doc As NotesDocument
'Object creation
    Set Workspace = New NotesUIWorkspace
    Set Db = Session.CurrentDatabase
'Get handle on the current document
    Set UIDoc = Workspace.CurrentDocument
    Set Doc = UIDoc.Document
'Set the View object
    Set View = Db.GetView("Employees")
'Check to see if the view object was set
    If Not (View Is Nothing) Then
'Get a handle on the parent document
        Set parentDoc = View.GetDocumentByKey("William
Thompson")
'Check to see if the parent document was found
        If Not (parentDoc Is Nothing) Then
'Make the current document a response document to parentDoc
            Call Doc.MakeResponse(parentDoc)
    Call Doc.Save(True, False)
        Else
            Messagebox "The parent document could not be
found."
        End If
    Else
        Messagebox "The view could not be opened."
```

```
        End If
        Exit Sub
ErrorHandler:
        Messagebox "There were problems with the response document
creation process."
End Sub
```

Processing Existing Documents

```
Sub Click(Source As Button)
    On Error Goto ErrorHandler
'Declarations
        Dim Workspace As NotesUIWorkspace
        Dim UIDoc As NotesUIDocument
        Dim Session As NotesSession
        Dim Db As NotesDatabase
        Dim View As NotesView
        Dim consultantView As NotesView
        Dim parentDoc As NotesDocument
        Dim Doc As NotesDocument
'Object creation
        Set Workspace = New NotesUIWorkspace
        Set Db = Session.CurrentDatabase
        Set consultantView = Db.GetView("AllStaffConsultants")
'Get handle on the current document
        Set UIDoc = Workspace.CurrentDocument
        Set Doc = UIDoc.Document
'Set the View object
        Set View = Db.GetView("Employees")
'Check to see if the view object was set
        If Not (View Is Nothing) Then
'Get a handle on the parent document
            Set parentDoc = View.GetDocumentByKey("William
Thompson")
'Check to see if the parent document was found
            If Not (parentDoc Is Nothing) Then
'Loop through the AllStaffConsultants view;make them responses
to the parentDoc object
                Set Doc = consultantView.GetFirstDocument
                Do While Not (Doc Is Nothing)
Call Doc.MakeResponse(parentDoc)
Call Doc.Save(True,False)
                    Set Doc =
consultantView.GetNextDocument(Doc)
    Loop
            Else
                Messagebox "The parent document could not be
found."
            End If
        Else
            Messagebox "The view could not be opened."
        End If
        Exit Sub
ErrorHandler:
        Messagebox "There were problems with the response document
creation process."
End Sub
```

This example shows the creation of multiple-response documents to one parent by looping through a view, which happens to contain all the documents we need to process.

Copying Documents

When building an application, you may come across a situation where you need to copy documents from one database to another —or make a copy of a document within the same database. The NotesDatabase class CopyToDatabase method will help you do this task. When copying a document to a database, you must have proper access to do so (at least Author Access).

Syntax
```
Set targetDocument = sourceDocument.CopyToDatabase(
targetDatabase )
```

Parameter
targetDatabase   The database in which you want a copy of the document.

Return Value
targetDocument   The new document in the specified targetDatabase.

Copying Documents Examples

Copying One Document within the Current Database
```
Sub Click(Source As Button)
'Declare the appropriate objects
    Dim Workspace As NotesUIWorkspace
    Dim UIDoc As NotesUIDocument
    Dim Session As NotesSession
    Dim TargetDb As NotesDatabase
    Dim SourceDoc As NotesDocument
    Dim TargetDoc As NotesDocument
'Set the objects
    Set Workspace = New NotesUIWorkspace
    Set UIDoc = Workspace.CurrentDocument
    Set SourceDoc = UIDoc.Document
    Set TargetDb = Session.CurrentDatabase
'Copy the SourceDoc to the TargetDoc to the TargetDb
    Set TargetDoc = SourceDoc.CopyToDatabase(TargetDb)
    Call TargetDoc.Save(True,False)
'Display prompt with the new Universal IDs
    Messagebox "This document ID: " & SourceDoc.UniversalID
    Messagebox "New document ID: " & TargetDoc.UniversalID
End Sub
```

This example shows how a handle is placed on the source document through the NotesUIDocument class CurrentDocument property. When copying a document within the same database, Notes will automatically assign the new document a new universal ID.

Copying Multiple Documents within the Current Database

```
Sub Click(Source As Button)
'Declare the appropriate objects
    Dim Session As NotesSession
    Dim SourceDb As NotesDatabase
    Dim SourceDbDocuments As NotesDocumentCollection
    Dim SourceDoc As NotesDocument
    Dim TargetDoc As NotesDocument
'Set the SourceDb object handle
    Set SourceDb = Session.CurrentDatabase
    Set SourceDbDocuments = SourceDb.AllDocuments
'Copy the SourceDb's documents to the TargetDb (which happens
to be the SourceDb)
    For j = 1 To SourceDbDocuments.Count
        Set SourceDoc = SourceDbDocuments.GetNthDocument(j)
        Set TargetDoc = SourceDoc.CopyToDatabase(SourceDb)
        Call TargetDoc.Save(True,False)
    Next
End Sub
```

Copying Multiple Documents to Another Database

```
Sub Click(Source As Button)
'Declare the appropriate objects
    Dim Session As NotesSession
    Dim SourceDb As NotesDatabase
    Dim SourceDbDocuments As NotesDocumentCollection
    Dim TargetDb As NotesDatabase
    Dim SourceDoc As NotesDocument
    Dim TargetDoc As NotesDocument
'Set the SourceDb object handle
    Set SourceDb = Session.CurrentDatabase
    Set SourceDbDocuments = SourceDb.AllDocuments
'Set the TargetDb object handle
    Set TargetDb = New NotesDatabase("","Archive.Nsf")
'Copy the SourceDb's documents to the TargetDb
    For j = 1 To SourceDbDocuments.Count
        Set SourceDoc = SourceDbDocuments.GetNthDocument(j)
        Set TargetDoc = SourceDoc.CopyToDatabase(TargetDb)
        Call TargetDoc.Save(True,False)
    Next
End Sub
```

This example makes use of a NotesDocumentCollection object and loops through the collection to copy all the documents from the source database to the target.

Deleting Documents

You can use the NotesDocument class Remove method to delete a document from a database. Once again, to do this, you must have sufficient access to the database and the document.

Syntax

```
flag = notesDocument.Remove( force )
```

Parameter

force **Boolean.** If True, the document will be deleted—even if another user modified the document since the script opened it. If False, the document will not be deleted if it has been modified by another user.

Return Value

True   The document was successfully deleted.

False   The document was not deleted. Another user modified it, and the force parameter is set to False.

Deleting Documents Example

```
Sub Click(Source As Button)
        On Error Goto ErrorHandler
'Declarations
        Dim Db As NotesDatabase
        Dim View As NotesView
        Dim Doc As NotesDocument
'Object creation
        Set Db = New NotesDatabase("","")
        openFlag = Db.OpenByReplicaID( "", "85255FA900747B84" )
'Check to see if the database was opened
        If openFlag = True Then
'Set the View object
            Set View = Db.GetView("Employees")
'Check to see if the view object was set
            If Not (View Is Nothing) Then
'Get a handle on the appropriate document
                Set doc = View.GetDocumentByKey("William
Thompson")
'Check to see if the document was found
                If Not (Doc Is Nothing) Then
'Delete the document
                    Call Doc.Remove(True)
                Else
                    Messagebox "There was an error deleting the
document."
                End If
            Else
                Messagebox "The view could not be opened."
            End If
```

```
      Else
            Messagebox "The database could not be open."
      End If
      Exit Sub
ErrorHandler:
      Messagebox "There were problems with the deletion
process."
End Sub
```

In this example, we use the NotesView class GetDocumentbyKey method to place a handle on the document we want to delete. Once again, note the data validation and error-handling.

Documents and Folders

Putting a Document in a Folder

To put a document in a folder, you must

- Identify the document for placement.
- Identify which folder into which you want the document to be placed (usually based on some field value).
- Use the NotesDocument class PutInFolder method.

NotesDocument Class PutInFolder Method

Syntax
```
      Call notesDocument.PutInFolder( folderName$ )
```

Parameter
folderName$ **String.**   The name of the folder in which to place the document. The folder may be personal if the script is running on the workstation. If the folder is within another folder, specify a path to it—separating folder names with backward slashes. For example: *Vehicles\Bikes*.

Usage
If the document is already inside the folder you specify, PutInFolder does nothing. If you specify a path to a folder, and none of the folders exists, the method creates all of them for you. If you do not have rights to create a new folder, your script will return an error.

Removing a Document from a Folder

To remove a document from a folder, you must

- Identify the document you want to remove from the folder
- Identify which folder you want the document to be removed from
- Use the NotesDocument class RemoveFromFolder method

NotesDocument Class RemoveFromFolder Method
Syntax

```
Call notesDocument.RemoveFromFolder( folderName$ )
```

Parameter

folderName$ **String.** The name of the folder from which to remove the document. The folder may be personal if the script is running on a workstation. If the folder is within another folder, specify a path to it, separating folder names with backward slashes, for example: *Vehicles\Bikes*.

Usage

The method does nothing if the document is not in the folder you specify, or if the folder you specify does not exist.

Documents and Folders Examples

Putting a Document in a Folder

```
Sub Click(Source As Button)
'Declarations
        Dim Workspace As NotesUIWorkspace
        Dim UIDoc As NotesUIDocument
        Dim Doc As NotesDocument
'Set objects
        Set Workspace = New NotesUIWorkspace
        Set UIDoc = Workspace.CurrentDocument
        Set Doc = UIDoc.Document
'Put the document in the "Employees" folder
        Call Doc.PutInFolder("Employees")
End Sub
```

This example places the current document inside the *Employees* folder.

Removing a Document from a Folder

```
Sub Click(Source As Button)
'Declarations
    Dim Workspace As NotesUIWorkspace
    Dim UIDoc As NotesUIDocument
    Dim Doc As NotesDocument
'Set objects
    Set Workspace = New NotesUIWorkspace
    Set UIDoc = Workspace.CurrentDocument
    Set Doc = UIDoc.Document
'Put the document in the "Employees" folder
    Call Doc.RemoveFromFolder("Employees")
End Sub
```

This example removes the current document from the *Employees* folder.

Sending Mail

Scripts can generate mail messages using the full capacities of Notes-Mail, including store with document, mail encryption, and signature. Sending a mail message is similar to creating a new document.

- Select an existing document, or create a new document

- Add any items (both items that hold data and those items that hold mail information, such as SendTo and Subject items)

- Instead of calling the NotesDocument class Save method, call the Send method. If you choose to create a new document to send, you do not need to save the document to send it (unless you want to).

Syntax

```
Call notesDocument.Send( attachForm [, recipients ] )
```

Parameters

attachForm **Boolean.** If True, the form is stored and is sent along with the document. If False, it is not.

recipients **String or array of strings.** Optional. The recipients of the document, which might include people, groups, or mail-in databases. This parameter is ignored if the document contains a SendTo item, in which case the document is mailed to recipients listed in SendTo. If the document does not contain a SendTo item, it is required.

There are two kinds of items that can affect the mailing of the document when you use Send:

- If the document contains additional recipient items, such as CopyTo or BlindCopyTo, Notes mails the documents to these recipients.

- If the document contains items to control the routing of mail, such as DeliveryPriority, DeliveryReport, or ReturnReceipt, Notes uses these items when sending the document.

The SaveMessageOnSend property controls whether the sent document is saved in the database. If SaveMessageOnSend is True and you attach the form to the document, the form is saved with the document.

Sending the form increases the size of the document, but it ensures that the recipient can see all of the items on the document. If a document is mailed by a script, the Send method automatically creates an item called $AssistMail on the sent document. The SentByAgent property uses this item to determine whether a document was mailed by a script.

If a script runs on a workstation, the mailed document contains the current user's name in the From item. If a script runs on a server, the mailed document contains the server's name in the From item.

Sending a Mail Example

```
Sub Click(Source As Button)
'Declarations
    Dim Session As NotesSession
    Dim Db As NotesDatabase
    Dim mailDoc As NotesDocument
    Dim rtItem As NotesRichTextItem
'Set objects
    Set Session = New NotesSession
    Set Db = Session.CurrentDatabase
    Set mailDoc = New NotesDocument(Db)
    Set rtItem = New NotesRichTextItem(mailDoc,"Body")
'Add items to mailDoc
    mailDoc.SendTo = "Sean Keighron"
    mailDoc.Subject = "New CD is out!"
'Format the Rich Text Body field
    Call rtItem.AppendText("Dear " & mailDoc.SendTo(0) & ",")
    Call rtItem.AddNewLine(2)
    Call rtItem.AppendText("You haven't registered for the new
CD! Please do it soon.")
    Call rtItem.AddNewLine(2)
    Call rtItem.AppendText("Thanks!")
    Call rtItem.AddNewLine(2)
```

```
        Call rtItem.AppendText(Session.CommonUserName)
   'Send the Document
        Call mailDoc.Send(False)
   End Sub
```

Notice that this example creates a rich text item on the document and adds text and formatting to it. If I wanted to save this document when I sent it, I could either call the NotesDocument class Save method, or I could have set the NotesDocument class SaveOnSend property to True.

Newsletters

A newsletter is a summary document containing information from and/or links to other documents. Newsletter objects are

- Built from a document pool
- Displayed one line per document, including a DocLink and descriptive field
- Contained in a NotesDocument object for saving or mailing

Figure 6.1 shows a Newsletter.

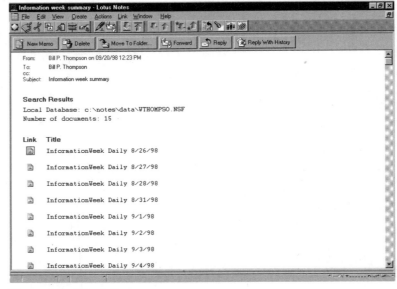

Figure 6.1 Newsletter

Making a Notes Newsletter

The essential steps for creating a newsletter are

1. Using the New keyword in the Dim statement to create a NotesNewsletter object
2. Calling one of the following NotesNewsletter methods to create a document

   - FormatDocument to create a NotesDocument object that contains a rendering of other documents
   - FormatMsgWithDoclinks to create a document that contains doclinks to other documents, plus header information

You can use the Send method of NotesDocument to mail the resulting newsletter.

Table 6.1 lists the NotesNewsletter properties.

Table 6.1   NotesNewsletter Properties

Property	Data type	Description
DoScore	Boolean	(Read-write) True if the newsletter includes each document's relevance score
DoSubject	Boolean	(Read-write) True if the newsletter includes each document's subject
SubjectItemName	String	(Read-write) Name of the field in each source document containing the subject. This is how NotesNewsletter tags each link

Creating a Newsletter Example

```
Sub Click(Source As Button)
'Declarations
    Dim Session As New NotesSession
    Dim CurrentDatabase As NotesDatabase
    Dim SourceDatabase As New NotesDatabase("","")
    Dim SourceCollection As NotesDocumentCollection
    Dim dateTime As NotesDateTime
```

```
'Get a handle on the current database
    Set  CurrentDatabase = Session.CurrentDatabase
'Get a handle on the database to search for a document pool
    Call SourceDatabase.Open("", "wthompso.NSF")
'Set the Search parameters
    Set dateTime = New NotesDateTime(SourceDatabase.Created)
    SearchFormula$ = "From = ""InformationWeek
<news@daily.informationweek.com>"""
'Search the Source Database for the document pool
    Set SourceCollection =
SourceDatabase.Search(SearchFormula$, dateTime, 0)
'Check to see if there are any documents in the
DocumentCollection. If not, exit the sub
    If (SourceCollection Is Nothing) Or
(SourceCollection.Count = 0) Then
        Messagebox "There are no messages from
InformationWeek."
        Exit Sub
    End If
'Create the newsletter document
    Set NewsLetterDocument = New
NotesNewsletter(SourceCollection)
'Add the newsletter parameters
    NewsLetterDocument.DoSubject = True
    NewsLetterDocument.DoScore = False
    NewsLetterDocument.SubjectItemName = "Subject"
'Create a new NotesDocument for mailing
    Dim MyDocument As New NotesDocument(CurrentDatabase)
    Set MyDocument=NewsLetterDocument.FormatMsgWithDocLinks
(CurrentDatabase)
'Add items for mailing
    MyDocument.Form = "Memo"
    MyDocument.Subject = "Information week summary"
    MyDocument.SendTo = Session.CommonUserName
'Send the document
    Call MyDocument.Send(False)
    Messagebox "Newsletter has been sent."
End Sub
```

Sample Questions

1. *Objective:* Creating a new NotesDocument

   A back-end method for creating a Notes document is

   a. Compose
   b. Create
   c. CreateDocument
   d. CreateNewDocument

2. *Objective:* Document creation

   When creating a document using a back-end class

   a. The document is saved when it is created.
   b. The document contains all items from the specified form.
   c. The document is not saved until the NotesDocument class SaveDocument method is called.
   d. The document is not saved until the NotesDocument class Save method is called.

3. *Objective:* Document creation

   The correct syntax for creating a NotesDocument object using the New method is

   a. Dim *variableName* As NotesDocument
   b. Dim *variableName* As New NotesDocument(*notesDatabase*)
   c. Dim *variableName* As New NotesDocument
   d. None of the above

4. *Objective:* Creating response documents

   When creating a response document, you must have a handle on

   a. The response document only
   b. A document collection and the parent document
   c. The parent document as well as the document that will become its response
   d. The parent document only

5. *Objective:* Notes documents and folders

   When using the PutInFolder method, if the folder does not exist

   a. You must write code to create the folder.
   b. You must trap the error.
   c. You must use the CreateFolder method to create the folder.
   d. The PutInFolder method will automatically create the folder.

6. *Objective:* Sending mail

   When using LotusScript to send mail

   a. You must save the document before sending.
   b. The document must already exist.
   c. If the document being sent does not have a SendTo item, you must add a SendTo item to the document or supply the recipients parameter for the Send method.
   d. The Send method attachForm parameter should always be set to True.

7. *Objective:* Sending mail

   If the document contains additional recipient items, such as CopyTo or BlindCopyTo

   a. The recipients listed in these items will also receive the document.
   b. Script will ignore the items.
   c. The recipients listed in these items will receive the document only if the IncludeAdditionalRecipients property is set.
   d. All of the above

8. *Objective:* Sending mail

   True or False: The SaveMessageOnSend property is the only way to control whether the sent document is saved in the database.

9. *Objective:* Newsletters

   The FormatDocument method

   a. Creates a NotesDocument object that contains a rendering of other documents with doclinks

   b. Formats the font of a newsletter document

   c. Creates a NotesDocument object that contains a rendering of other documents

   d. All of the above

10. *Objective:* Newsletters

    The DoScore property of the NotesNewsletter class

    a. Is a read-only property and cannot be set

    b. Includes each document's relevance score if set to True

    c. Includes a doclink for each document in the Newsletter if set to True

    d. There is no DoScore property

11. *Objective:* Mailing documents

    Sean wants to create a new document and mail it to his friend. What is wrong with this script?

    ```
    . . . Declarations . . .
      Set Db = Session.CurrentDatabase
      Set Doc = New NotesDocument(Db)
      Doc.SendTo = "George Thomson"
      Doc.Subject = "Late Reminder"
      Call Doc.Save(True,False)
    ```

    a. There is no Body field for a message to the recipient.

    b. To create a new document, you must use the CreateDocument method.

    c. The document was never sent.

    d. This script is correct.

12. *Objective:* Creating response documents

    Amy wants to create a response document for each document
    that is in the NotesDocumentCollection. Which line causes
    Amy's intent to fail?

    ```
    . . . Declarations . . .
        Set Collection = Db.AllDocuments
        For j = 1 To Collection.Count
            Set Doc = Collection.GetNthDocument(j) - Line 1
            Set ResponseDoc = New NotesDocument(Db) - Line 2
            Call ResponseDoc.Save(True,False) - Line 3
            Call ResponseDoc.MakeResponse(Doc) - Line 4
        Next
    ```

    a. Line 1
    b. Line 2
    c. Line 3
    d. Line 4

13. *Objective:* Creating response documents

    Michael wants to send out a newsletter that contains links to
    the search results. Which line will cause the script to fail?

    ```
    . . . Declarations . . .
        Set SourceCollection =
    SourceDatabase.Search(SearchFormula$, dateTime, 0)
        Set NewsLetterDocument = New NotesNewsletter-Line 1
        NewsLetterDocument.DoSubject = True-Line 2
        NewsLetterDocument.DoScore = False
        NewsLetterDocument.SubjectItemName = "Subject"-Line 3
        Dim MyDocument As New NotesDocument(CurrentDatabase)
        Set MyDocument=NewsLetterDocument.FormatMsgWithDocLinks
    (CurrentDatabase)-Line 4
        MyDocument.Form = "Memo"
        MyDocument.Subject = "Information week summary"
        MyDocument.SendTo = Session.CommonUserName
        Call MyDocument.Send(False)
    ```

    a. Line 1
    b. Line 2
    c. Line 3
    d. Line 4

Sample Answers

1. *Answer:* c

   Use the New method of a NotesDocument class or the CreateDocument method from the NotesDatabase class to create a new document in a database. Both methods will create a new instance of a NotesDocument object reference.

2. *Answer:* d

   You use the Save method of the NotesDocument class to save a document to disk. If you do not use the Save method, the new document is lost when the program exits. When creating a NotesDocument object, you do not need to specify a form as a parameter to create it. If you add a NotesItem called *form* and populate it with a form name, it does not mean that all the items for that form will be placed on your document. Remember, the form design element and the data that show through the form are separate.

3. *Answer:* b

   ```
   Dim variableName As New NotesDocument( notesDatabase ).
   ```

4. *Answer:* a

   Call notesDocument.MakeResponse(parentDocument) is the syntax for making a document a response. You must have a handle on the parent document—and either a newly created NotesDocument object or an existing Notes document.

5. *Answer:* d

   When using the PutInFolder method, if the folder you want to put your document into does not exist, it will automatically create the folder using the default folder view as its design template.

6. *Answer:* c

   A NotesDocument object does not need to be saved before mailing. Just as in your mailbox or when you are using @MailSend, if you do not specify a recipient for you mail message, Notes will return an error. With LotusScript, if you do not have a SendTo item on your NotesDocument, you must either create one or supply the recipients parameter with some names.

7. *Answer:* d

   If the document contains additional recipient items, such as CopyTo or BlindCopyTo, Notes mails the documents to these recipients. No property exists called IncludeAdditionalRecipients.

8. *Answer:* False

   The SaveMessageOnSend property controls whether the sent document is saved in the database but you can also call the NotesDocument class Save method to save the document as well without setting this property.

9. *Answer:* d

   The FormatDocument method creates a NotesDocument object that contains a rendering of other documents.

10. *Answer:* b

    The DoScore property is a read/write property which, if set to True, includes each document's relevance score.

11. *Answer:* c

    The script is incorrect, because the document was never sent. You must use the NotesDocument Send method to send the document.

12. *Answer:* d

    Line 4 causes the script to fail, because you must call the MakeResponse method before you save the document. Otherwise, that change will not take effect.

13. *Answer:* d

    Line 1 causes the script to fail, because it is not the correct syntax for creating a NotesNewsLetter object. The line should read

```
Set NewsLetterDocument = New NotesNewsletter(SourceCollection)
```

CHAPTER 7

Processing Document Items

This chapter reviews processing document items. The NotesItem and NotesRichTextItem classes enable you to examine and manipulate item properties and contents. You gain access to a NotesItem or NotesRichTextItem object through various methods in the NotesDocument class. The NotesRichTextItem class inherits from NotesItem, meaning that NotesRichTextItem objects can use all the properties and methods of NotesItem. You can

- Get an item and its values
- Create an item and assign values
- Copy an item
- Remove an item
- Access item properties
- Work with a rich text item
- Work with an embedded object

Chapter Objectives

The objectives of this chapter are to increase your understanding of the following items:

- Creating items
- Copying items
- Deleting items
- Retrieving item values
- Assigning item values
- Items and multi-value fields
- Rich text items

Creating, Setting, and Removing Items

Getting Item Values

The NotesDocument class provides several properties and methods to access the items in the document. If you do not know the name of an item or want to traverse the items in a document, you can use the Items property, which is an array of NotesItem objects. As you access each array element, use the NotesItem properties and methods to retrieve and set the values of the object.

To access an item whose name is known, use the GetFirstItem method of the NotesDocument class.

You can get the value of an item through several techniques.

 NOTE
The first two techniques do not require that you first access the item object.

- After accessing the NotesDocument object containing the item, you can specify the item name as though it were a property of the document. For example, if *doc* were the name of a document object and Subject is an item in the document, doc.Subject is the value of the item. This syntax is referred to as the *extended syntax*.
- You can use the GetItemValue method of NotesDocument.
- You can access the item and examine its Values property.

 NOTE
Two items can have the same name. If you need to access an item other than the first item, use the Items property of NotesDocument and then loop through the items until you find what you need.

If you need to loop through items, if the item is

- Not a rich text field and not a multi-value field. You can always deal with only the (0) element in the array.

- A multi-value field. You must anticipate any number of array elements.

- Not rich text but programmatically was given multiple items with the same name. Using the NotesDocument object AppendItemValue method, you must anticipate all items using the GetFirstItem/GetnNextItem methods.

- Rich text and is larger than 64KB. You have to anticipate multiple items all with the same name.

GetItemValue Method

Syntax
```
valueArray = notesDocument.GetItemValue( itemName$ )
```

Parameter

itemName$ **String.**   The name of an item.

Return Value

value   The value or values contained in itemName$. The data type of the value depends on the data type of the item.

Table 7.1 shows several NotesItem types and their value return types.

Table 7.1   NotesItem Types

NotesItem Type	Value Return Type
Rich text	String. The text in the item, rendered into plain text
Text or text list (includes Names, Authors, and Readers item types)	Array of strings
Number or number list	Array of doubles
Date-time or range of date-time values	Array of doubles

When GetItemValue returns an array, each element in the array corresponds to a value in the item. If the item contains a single value, the array has just one element.

For text, number, and time-date items, *GetItemValue always returns an array,* even when there is only a single value in the item. If you know the item contains only a single value, access the first element in the array, which is at index zero. If you know the item contains multiple values, but you do not know how many, iterate over the array elements using the Forall statement.

Values Property

Read/Write. The value(s) that an item holds.

To get: valueArray = notesItem.Values

To set: notesItem.Values = valueArray

For text, number, date-time items, and attachments. This property always returns an array, even when there is only a single value in the item. If you know the item contains only a single value, access the first element in the array, which is at index 0. If you know the item contains multiple values, but you do not know how many, iterate over the array using the Forall statement.

- For attachments, the attachment name can be passed to the GetAttachment method in the NotesDocument.

- This property returns the same value(s) for an item as the GetItemValue method in the NotesDocument.

Table 7.2 shows some item types and their values data types.

Table 7.2 NotesItem Types

Item Type	Values Data Type
Rich text	String. The text in the field, rendered into plain text
Text or text list (includes Names, Authors, and Readers item types)	Array of strings

Item Type	Values Data Type
Number or number list	Array of doubles
Date-time or range of date-time values	Array of doubles
Attachment	Array of strings. The name of the attachment in element zero

Extended Syntax

The extended syntax enables you to refer to an item as though it were a property of the document object (e.g., Document.ItemName). Remember that items return an array of values. If you anticipate a multi-value field you should create a variant to hold the values and loop through them. Using the extended syntax shortens the amount of coding you may need to set field values. Here are a few iterations of the way you can use the extended syntax:

To get a singular value: variableName = NotesDocument.itemName(0)

To get multi values: variant = NotesDocument.itemName

 TIP
Most of the questions on the exam that have coding examples use the extended syntax.

Getting Item Values Examples

```
Sub Initialize
'Declarations
    Dim Session As NotesSession
    Dim Db As NotesDatabase
    Dim View As NotesView
    Dim Doc As NotesDocument
'Set Objects
    Set Session = New NotesSession
    Set Db = Session.CurrentDatabase
    Set View = Set View = Db.GetView("EmployeeNamesByTitle")
    Set Doc = View.GetDocumentByKey("Manager")
'Use the GetItemValue method to get the employee's name
    x = Doc.GetItemValue("EmployeeName")
End Sub
```

Once again, the value returned is of a variant data type—because the variable name is not declared.

Values Property Example

```
Sub Initialize
'Declarations
     Dim Session As NotesSession
     Dim Db As NotesDatabase
     Dim View As NotesView
     Dim Doc As NotesDocument
     Dim Item As NotesItem
'Set Objects
     Set Session = New NotesSession
     Set Db = Session.CurrentDatabase
     Set View = Db.GetView("EmployeeNamesByTitle")
     Set Doc = View.GetDocumentByKey("Manager")
'Use the GetFirstItem method to get a handle on the
"EmployeeName" item
     Set Item = Doc.GetFirstItem("EmployeeName")
'Use the Values property to get the values in the EmployeeName
item
     EmployeeName = Item.Values
End Sub
```

In this script example, a NotesItem object was declared so the Values property could be accessed.

Extended Syntax Example

```
Sub Initialize
'Declarations
     Dim Session As NotesSession
     Dim Db As NotesDatabase
     Dim View As NotesView
     Dim Doc As NotesDocument
'Set Objects
     Set Session = New NotesSession
     Set Db = Session.CurrentDatabase
     Set View = Set View = Db.GetView("EmployeeNamesByTitle")
     Set Doc = View.GetDocumentByKey("Manager")
'Use the extended syntax to get the employee's name
     x = Doc.EmployeeName(0)
End Sub
```

In this example, the script gets the value of the employee's name by using the extended syntax. Because x is not declared, the value that is returned from the extended syntax will be a variant data type.

Creating Items and Assigning Values

The New method of the NotesItem class creates a new item in a document and assigns a value to it. You can also use the AppendItemValue method of the NotesDocument class to create a new item. If another item exists with the same name, the new item does not replace that item—but rather becomes an additional item.

To assign and reassign values to an item:

- You can write to the item name as though it were a property of the document. For example, if *doc* is the name of a document object, and *Subject* is an item in the document, you can write to *doc.Subject*. The assigned value can be any LotusScript data type, scalar or array, a reference to a NotesItem object, a reference to a NotesDateTime object, or a reference to a NotesDateRange object.

- You can use the ReplaceItemValue method of the NotesDocument, which is equivalent to the first technique. If the item does not exist, ReplaceItemValue will create it.

- You can access the item and write to its Values property.

Writing to the Values property affects only the current item object, while the other techniques replace all values of all items with the specified name.

After creating or replacing an item, you must call the Save method for the NotesDocument object containing the item—or the update is lost when the program exits.

Computed fields are not recalculated when you save a document. If you change items that affect a computed field and want the new value to be available immediately, you must calculate the value in the script. For example, you should do this if you want the user to see the new value in a view—or you want to use the new value in a calculation.

 NOTE
The user cannot see the new item as a field in the document unless you add the field to the form design. The item exists, however, and can be accessed programmatically.

An item cannot appear in a view unless the IsSummary property is True. IsSummary is always False for rich text items, which cannot appear in views. Also, IsSummary is initially False if the item is created with New or ReplaceItemValue.

You can replace the value of an item in every document of a collection with the StampAll method of NotesDocumentCollection. This method writes immediately to the back-end documents. You do not have to call the NotesDocument class Save method after using StampAll.

NotesItem New Method

Syntax

Given a document, New creates an item on the document with a name and value that you specify. The data type of the item depends on the value you give it.

You must call Save on the document if you want the modified document to be saved to disk. The document will not display the new item in the user interface unless there is a field of the same name on the form used to display the document.

```
Dim variableName As New NotesItem( notesDocument, name$, value
[, specialType% ] )
```

or

```
Set notesItem = New NotesItem( notesDocument, name$, value  [,
specialType% ] )
```

Parameters

notesDocument   The document on which to create the item.

name$ **String.**   The name of the new item.

value   The value to assign to the new item. The data type of *value* determines the type of item that Notes creates, as seen in Table 7.3.

Table 7.3   Data Type and NotesItem Type

Value Data Type	NotesItem Type
String	Text (if *specialType%* is used, may be Names, Readers, or Authors).
Array of strings	Text (if *specialType%* is used, may be Names, Readers, or Authors).
Integer, long, single, double, currency	Number
Array of integers, longs, singles, doubles, or currencies	Number
Variant of type DATE	Time
Array of variants of type DATE	Time

`specialType%` **Optional.**   Constant of type integer. Indicates whether a text item should be of type Names, Readers, or Authors. Must be one of the following constants: NAMES, READERS, or AUTHORS. To use *specialType%*, the *value* parameter must be a string or array of strings.

AppendItemValue Method
Syntax
Creates a new item on a document and sets the item value.

```
Set notesItem = notesDocument.AppendItemValue( itemName$, value )
```

Parameters
`itemName$` **String.**   The name of the new item.

`value`   The value of the new item. The data type of the new item depends on the data type of the *value* that you place in it. Table 7.4 shows several data types of value and their resulting NotesItems.

Table 7.4   Data Type and NotesItem Type

Data Type of Value	Resulting NotesItem
String	Text item containing *value*
Array of strings	Text item containing each element of *value*
Integer	Number item containing *value*
Array of integers	Number item containing each element of *value*
Variant of type DATE	Date-time item containing *value*
Array of variants, where each variant is of type DATE	Date-time item containing each element of *value*
NotesDateTime	Date-time item containing the date-time represented by the object
NotesItem	Item whose data type matches the NotesItem type and whose value(s) match the NotesItem value(s)

Return Value

`notesItem`   The new item. To keep the new item in the document, you must call the Save method after calling AppendItemValue.

If the document already has an item called *itemName$,* Append-ItemValue does not replace it. Instead, it creates another item of the same name and gives it the value you specify.

The IsSummary property of the new item defaults to True, which means that the item value can be displayed in a view or folder.

ReplaceItemValue Method

Replaces all items of the specified name with one new item, which is assigned the specified value. If the document does not contain an item with the specified name, the method creates a new item and adds it to the document.

Syntax

```
Set notesItem = notesDocument.ReplaceItemValue( itemName$,
value )
```

Parameters

`itemName$` **String.**   The name of the item(s) you want to replace.

`value`   The value of the new item. The data type of the item depends on the data type of *value* and does not need to match the data type of the old item.

Table 7.5 shows several data types of value and their resulting NotesItems.

Table 7.5   Data Type and NotesItem Type

Data Type of Value	Resulting NotesItem
String	Text item containing *value*
Array of strings	Text item containing each element of *value*
Integer	Number item containing *value*
Array of integers	Number item containing each element of *value*
Variant of type DATE	Time-date item containing *value*
Array of variants, where each variant is of type DATE	Date-time item containing each element of *value*

Data Type of Value	Resulting NotesItem
NotesTimeDate	Date-time item containing the date-time represented by the object
NotesItem	Item whose data type matches the NotesItem type and whose value(s) match the NotesItem value(s)

Return Value

`notesItem` The new item, which replaces all previous items that had *itemName$*. The new item has a new *value,* which may be of a different data type than the old item.

> **NOTE**
> The IsSummary property of an item defaults to False, which means that the item value cannot be displayed in a view or folder. You must explicitly set IsSummary to True if you want the value to be displayed in a view or folder.

Extended Syntax

The extended syntax enables you to refer to an item as though it were a property of the document object (e.g., Document.ItemName). Remember that items return an array of values. If you anticipate a multi-value field, you should create a variant to hold the values and loop through them. Using the extended syntax shortens the amount of coding you may need to set field values. Here are a few iterations of the way you can use the extended syntax:

To set a singular value: NotesDocument.itemName = variableName or literal value

To set multi-values: NotesDocument.itemName = variant or array

Creating and Assigning Values Examples

NotesItem New Method Example

```
Sub Initialize
'Declarations
    Dim Session As NotesSession
    Dim Db As NotesDatabase
    Dim View As NotesView
    Dim Doc As NotesDocument
    Dim Item As NotesItem
```

```
'Set Objects
    Set Session = New NotesSession
    Set Db = Session.CurrentDatabase
    Set View = Db.GetView("EmployeeNamesByTitle")
    Set Doc = View.GetDocumentByKey("Manager")
'Create a new item on the document
    Set Item = New NotesItem(Doc, "Authors",
"[Administrators]", AUTHORS)
'Save the document so the new item saves to it
    Call Doc.Save(True,False)
End Sub
```

This script searches for a document and then uses the New method to add a NotesItem to the document. Notice the parameters supplied to the New method. The New method creates an item called Authors with a value of *[Administrators]* and of data type text with Summary Read/Write access (Authors data type).

AppendItemValue Method Example

```
Sub Initialize
'Declarations
    Dim Session As NotesSession
    Dim Db As NotesDatabase
    Dim View As NotesView
    Dim Doc As NotesDocument
    Dim Item As NotesItem
'Set Objects
    Set Session = New NotesSession
    Set Db = Session.CurrentDatabase
    Set View = Db.GetView("EmployeeNamesByTitle")
    Set Doc = View.GetDocumentByKey("Manager")
'Create a new item on the document
    Set Item =
Doc.AppendItemValue("EmployeeName",Session.UserName)
'Save the document so the new item saves to it
    Call Doc.Save(True,False)
End Sub
```

This script uses the AppendItemValue method to add an item called EmployeeName that will hold the value of the current user's name. Notice that the return value is a NotesItem object. If the document already has an item called EmployeeName, AppendItemValue does not replace it. Instead, it creates another item of the same name and gives it the value you specify.

ReplaceItemValue Method Example

```
Sub Initialize
'Declarations
      Dim Session As NotesSession
      Dim Db As NotesDatabase
      Dim View As NotesView
      Dim Doc As NotesDocument
      Dim Item As NotesItem
'Set Objects
      Set Session = New NotesSession
      Set Db = Session.CurrentDatabase
      Set View = Db.GetView("EmployeeNamesByTitle")
      Set Doc = View.GetDocumentByKey("Manager")
'Create a new item on the document
      Set Item =
Doc.ReplaceItemValue("EmployeeName",Session.UserName)
'Save the document so the new item saves to it
      Call Doc.Save(True,False)
End Sub
```

This script uses the ReplaceItemValue method to add an item called EmployeeName that will hold the value of the current user's name. Notice that the return value is a NotesItem object. If the document already has an item called EmployeeName, ReplaceItemValue will replace all items of the specified name with one new item, which is assigned the specified value.

Extended Syntax Example

```
Sub Initialize
'Declarations
      Dim Session As NotesSession
      Dim Db As NotesDatabase
      Dim View As NotesView
      Dim Doc As NotesDocument
'Set Objects
      Set Session = New NotesSession
      Set Db = Session.CurrentDatabase
      Set View = Db.GetView("EmployeeNamesByTitle")
      Set Doc = View.GetDocumentByKey("Manager")
'Create a new item on the document
      Doc.EmployeeName = Session.CommonUserName
'Save the document
      Call Doc.Save(True,False)
End Sub
```

This script uses the extended syntax to add an item called Employee-Name and assigns it the value of the current user's name. If the document already has an item called EmployeeName, the extended syntax will replace the item.

Items and Multi-Value Fields

To set multi-value fields, you must set the field with an array of values. The ability to read or set variables or item values with respect to arrays takes a bit of practice. This table outlines some common interactions between single-value variables and arrays and single-value fields and multi-value fields during variable assignment.

TIP
Understanding the relationship between items and multi-value fields is important for the exam. You should pay attention to the types of errors that result from incorrect references.

Table 7.6 shows several assignments and their results.

Table 7.6 Array Assignments

Assignment	Result
MyString = MyArray	Type-mismatch (compile error)
MyArray = MyString	Illegal reference to array (compile error)
MyString = MyArray(0)	OK
MyArray(0) = MyString	OK
MyDoc.SVField = MyArray	MyArray = [0]"Red";[1]"Green",[2]"Blue" MVField is assigned "Red;Green;Blue"
MyArray = MyDoc.MVField	MyArray = [0]"Red";[1]"Green",[2]"Blue" MVField is assigned "Red":"Green":"Blue"
MyArray = MyDoc.MVField	Illegal reference to array (compile error)
MyArray(0) = MyDoc.MVField	Type-mismatch
MyArray(0) = MyDoc.MVField(0)	OK

Assignment	Result
MyVariant = MyDoc.MVField	MVField = "Red":"Green":"Blue" MyVariant is assigned [0]"Red";[1]"Green",[2]"Blue"
MyVariant = MyArray	OK
MyArray = MyOtherArray	Illegal reference to array (compile error)
MyArray(0) = MyDoc.SVField(0)	OK
MyVariant = MyDoc.SVField	OK

Copying Items

There are three methods you can use to copy items:

- Use the CopyItemToDocument method of the NotesItem class to copy the current item to another document

- Use the CopyAllItems method of the NotesDocument to copy all items in the current document to another document

- Use the CopyItem method of NotesDocument to copy an item to another item in the same document

After copying an item, you must call the Save method for the NotesDocument object containing the new item, or the update is lost when the program exits.

CopyItemToDocument Method
Syntax

```
Set notesItem = notesItem.CopyItemToDocument( notesDocument,
newName$ )
```

Parameters

notesDocument   The document on which to create the item.

newName$ **String.**   The name of the new item. Specify an empty string (" ") if you want to keep the name of the original item.

Return Value

notesItem   The new item in the specified notesDocument, with a newName$.

When you call this method using a NotesRichTextItem object, file attachments, embedded objects, and object links that are contained within the rich text item are not copied to the destination document.

CopyAllItems Method
Syntax
```
Call notesDocument.CopyAllItems( notesDocument  [, replace] )
```

Parameters
`notesDocument`   The destination document.

`replace` **Boolean.**   If True, the items in the destination document are replaced. If False (default), the items in the destination document are added.

CopyItem Method
Syntax
```
Set notesItem = notesDocument.CopyItem( notesItem, newName$ )
```

Parameters
`notesItem`   The item you want to copy, usually from another document.

`newName$` **String.**   The name to assign to the copied item. Specify an empty string (" ") to retain the item's existing name.

Return Value
`notesItem`   A copy of the specified notesItem parameter, identical except for its newName$.

Copying Items Examples
CopyItemToDocument Example
```
Sub Initialize
'Declarations
     Dim Session As NotesSession
     Dim Db As NotesDatabase
     Dim View As NotesView
     Dim Doc As NotesDocument
     Dim NewDoc As NotesDocument
     Dim Item As NotesItem
     Dim NewItem As NotesItem
'Set Objects
     Set Session = New NotesSession
     Set Db = Session.CurrentDatabase
     Set View = Db.GetView("EmployeeNamesByTitle")
     Set Doc = View.GetDocumentByKey("Manager")
     Set Item = Doc.GetFirstItem("Employee")
```

```
'Create a new document
     Set NewDoc = New NotesDocument(Db)
'Use the CopyItemToDocument method
     Set NewItem = Item.CopyItemToDocument( Doc, "EmployeeName"
)
'Set other fields
     NewDoc.Form = "Employee Record"
     NewDoc.SendTo = Session.CommonUserName
     NewDoc.Subject = "Employee Record"
     Call NewDoc.Send(True)
'Save the document
     Call Doc.Save(True,False)
End Sub
```

CopyAllItems Example

```
Sub Initialize
'Declarations
     Dim Session As NotesSession
     Dim Db As NotesDatabase
     Dim View As NotesView
     Dim Doc As NotesDocument
     Dim NewDoc As NotesDocument
     Dim Item As NotesItem
     Dim NewItem As NotesItem
'Set Objects
     Set Session = New NotesSession
     Set Db = Session.CurrentDatabase
     Set View = Db.GetView("EmployeeNamesByTitle")
     Set Doc = View.GetDocumentByKey("Manager")
     Set Item = Doc.GetFirstItem("Employee")
'Create a new document
     Set NewDoc = New NotesDocument(Db)
'Use the CopyAllItems method
     Call Doc.CopyAllItems( NewDoc, True)
'Set other fields
     NewDoc.Form = "Employee Record"
     NewDoc.SendTo = Session.CommonUserName
     NewDoc.Subject = "Employee Record"
     Call NewDoc.Send(True)
'Save the document
     Call Doc.Save(True,False)
End Sub
```

CopyItem Example

```
Sub Initialize
'Declarations
     Dim Session As NotesSession
     Dim Db As NotesDatabase
     Dim View As NotesView
     Dim Doc As NotesDocument
     Dim NewDoc As NotesDocument
     Dim Item As NotesItem
     Dim NewItem As NotesItem
'Set Objects
     Set Session = New NotesSession
```

```
      Set Db = Session.CurrentDatabase
      Set View = Db.GetView("EmployeeNamesByTitle")
      Set Doc = View.GetDocumentByKey("Manager")
      Set Item = Doc.GetFirstItem("Employee")
'Create a new document
      Set NewDoc = New NotesDocument(Db)
'Use the CopyItem method
      Set NewItem = Doc.CopyItem( Item, "EmployeeName" )
'Set other fields
      NewDoc.Form = "Employee Record"
      NewDoc.SendTo = Session.CommonUserName
      NewDoc.Subject = "Employee Record"
      Call NewDoc.Send(True)
'Save the document
      Call Doc.Save(True,False)
End Sub
```

Removing Items

To remove items from a document, use either the Remove method of
the NotesItem class or the RemoveItem method of the Notes-
Document class. The Remove method removes only the current
object (other items with the same name remain). If there are multiple items with the same name, remove them all by looping through
them). The RemoveItem method removes all items with the specified
name. After removing an item, you must call the Save method for
the NotesDocument object that contained the item—or the update is
lost when the program exits.

Remove Method
Syntax
Permanently deletes an item from a document.

```
Call notesItem.Remove
```

RemoveItem Method
Given the name of an item, deletes the item from a document.

Syntax
```
Call notesDocument.RemoveItem( itemName$ )
```

Parameter
itemName$ **String.**   The name of the item to delete from the docu-
ment. If more than one item has *itemName$*, all items with this
name are deleted. If there is no item with *itemName$,* the method
does nothing.

Removing Items Examples

Remove Example

```
Sub Initialize
'Declarations
        Dim Session As NotesSession
        Dim Db As NotesDatabase
        Dim View As NotesView
        Dim Doc As NotesDocument
        Dim Item As NotesItem
'Set Objects
        Set Session = New NotesSession
        Set Db = Session.CurrentDatabase
        Set View = Db.GetView("EmployeeNamesByTitle")
        Set Doc = View.GetDocumentByKey("Manager")
        Set Item = Doc.GetFirstItem("Employee")
'Remove the item
        Call Item.Remove
'Save the document
        Call Doc.Save(True,False)
End Sub
```

RemoveItem Example

```
Sub Initialize
'Declarations
        Dim Session As NotesSession
        Dim Db As NotesDatabase
        Dim View As NotesView
        Dim Doc As NotesDocument
        Dim Item As NotesItem
'Set Objects
        Set Session = New NotesSession
        Set Db = Session.CurrentDatabase
        Set View = Db.GetView("EmployeeNamesByTitle")
        Set Doc = View.GetDocumentByKey("Manager")
'Remove the item
        Call Doc.Removetem("Employee")
'Save the document
        Call Doc.Save(True,False)
End Sub
```

Rich Text Items

Rich text items consist of multiple-linked items of the same name but are different from the following:

- A multi-value field, which is a single item that contains multiple values

- A non-rich text field that contains multiple items of the same name

Although NotesRichTextItem objects inherit all the properties of a NotesItem object, they have the additional capacity to contain compound document elements—including embedded and attached objects. Once you set an item to be rich text, you must use the NotesRichTextItem class methods to append text. You cannot use NotesItem methods to do so.

Operating with rich text fields is not as *visible* as working with non-rich text fields. You generally cannot see the effects of changes made by LotusScript until you save and open the document.

TROUBLESHOOTING TIP

When a rich text item is the return value of a method, such as GetFirstItem in NotesDocument, do not declare it with a Dim statement, but leave it a variant. It cannot be declared as a NotesRichTextItem object, because a *Type-mismatch* error occurs. If you declare it as a NotesItem object, you cannot use the NotesRichTextItem property and methods. A *Not a member* error occurs.

Creating a Rich Text Item

There are two ways to create a rich text item:

- Use the NotesDocument class CreateRichTextItem method
- Use the NotesRichTextItem class New method

CreateRichTextItem Method
Syntax
The CreateRichTextItem method creates a new rich text item on a document using a name you specify—and returns the corresponding NotesRichTextItem object. When used with OLE automation, this method enables you to create a new rich text item and NotesRichTextItem object without using New.

```
Set notesRichTextItem = notesDocument.CreateRichTextItem( name$ )
```

Parameter
name$ **String.** The name of the new rich text item.

Return Value
notesRichTextItem The newly created item.

NotesRichTextItem New Method

Given a document, New creates a rich text item on the document with the name you specify.

Syntax

```
Dim variableName As New NotesRichTextItem( notesDocument ,
name$ )
```

or

```
Set rotesRichTextItem = New NotesRichTextItem( notesDocument ,
name$ )
```

Parameters

notesDocument   The document in which to create a new rich text item.

name$ **String.**   The name of the new rich text item.

 NOTE
Because the NotesRichTextItem class inherits from the NotesItem class, all of the NotesItem properties and methods also can be used on a NotesRichTextItem. When you change the value of a NotesRichTextItem object, the change is not written to disk until you call the Save method for the parent NotesDocument.

Creating CreatingRichTextItem Examples

```
Sub Initialize
'Declarations
     Dim Session As NotesSession
     Dim Db As NotesDatabase
     Dim Doc As NotesDocument
     Dim rtItem As NotesRichTextItem
     Dim SendToArray(1)
'Set Objects
     Set Session = New NotesSession
     Set Db = Session.CurrentDatabase
     Set Doc = New NotesDocument(Db)
     SendToArray(0) = "Michael Butler"
     SendToArray(1) = "Larry Rosenbaum"
'Add Items to document
     Doc.Form = "Memo"
     Doc.Subject = "Rich text example"
     Doc.SendTo = SendToArray
'Create rich text item
     Call Doc.CreateRichTextItem("Body")
     Set rtItem = Doc.GetFirstItem("Body")
'Add text to the rich text item
     Call rtItem.AppendText("Here is my first line of text!")
```

```
      Call rtItem.AddNewLine(2)
      Call rtItem.AppendText("Here is my second line of text!!")
      Call rtItem.AddNewLine(2)
      Call rtItem.AppendText("Here is my third line of text!  A
virtual cornucopia of rich text!!!!")
'Send the document
      Call Doc.Send(False, SendToArray)
End Sub
```

NotesRichTextItem New Example

```
Sub Initialize
'Declarations
      Dim Session As NotesSession
      Dim Db As NotesDatabase
      Dim Doc As NotesDocument
      Dim rtItem As NotesRichTextItem
      Dim SendToArray(1) As Text
'Set Objects
      Set Session = New NotesSession
      Set Db = Session.CurrentDatabase
      Set Doc = New NotesDocument(Db)
      SendToArray(0) = "Michael Butler"
      SendToArray(1) = "Larry Rosenbaum"
'Add Items to document
      Doc.Form = "Memo"
      Doc.Subject = "Rich text example"
      Doc.SendTo = SendToArray
'Create rich text item
      Set rtItem = New NotesRichTextItem(Doc,"Body")
'Add text to the rich text item
      Call rtItem.AppendText("Here is my first line of text!")
      Call rtItem.AddNewLine(2)
      Call rtItem.AppendText("Here is my second line of text!
How prestidigitous!!!!")
      Call rtItem.AddNewLine(2)
      Call rtItem.AppendText("Here is my third line of text!  A
virtual cornucopia of rich text!!!!")
'Send the document
      Call Doc.Send(False, SendToArray)
End Sub
```

Creating Embedded Objects

The NotesEmbeddedObject class represents any one of the following items:

- An embedded object
- An object link
- A file attachment

The EmbeddedObjects property of NotesRichTextItem is an array of all the embedded objects, linked objects, and file attachments in a rich text item, which are of the NotesEmbeddedObject class. The HasEmbedded property of the NotesDocument tells you whether a document contains any embedded objects.

To get all the embedded objects in a document, including objects that were originally embedded in the form from which the document was created, use the EmbeddedObjects property of the Notes-Document. (This property does not include file attachments.)

To locate pre-r4 file attachments not associated with a rich text item, for example, get all the items in the document. For example, use the Items property in the NotesDocument. Test the Type property of each item for ATTACHMENT. To get an attachment of this type, use the GetAttachment method in the NotesDocument.

 NOTE
This feature is not supported under OS/2, UNIX, and on the Macintosh.

To create a new object, object link, or file attachment, use the EmbedObject method in the NotesRichTextItem class.

Syntax

```
Set notesEmbeddedObject = notesRichTextItem.EmbedObject( type%,
class$, source$, [ name$ ] )
```

Parameters

type% **Constant.** Indicates whether you want to create an attachment, an embedded object, or an object link. May be any of the following:

```
EMBED_ATTACHMENT
EMBED_OBJECT
EMBED_OBJECTLINK
```

class$ **String.** If you are using EMBED_OBJECT and want to create an empty embedded object from an application, use this parameter to specify the name of the application (for example, *123Worksheet*), and specify an empty string (" ") for source$. Case-sensitive.

If you are using EMBED_OBJECTLINK or EMBED_ATTACHMENT, specify an empty string (" ").

source$ **String.** If you are using EMBED_OBJECT and want to create an embedded object from a file, use this parameter to specify the name of the file, and specify an empty string (" ") for class$.

If you are using EMBED_ATTACHMENT or EMBED_OBJECTLINK, use this parameter to specify the name of the file to attach or link.

name$ **String.** Optional. Name by which you later can reference the NotesEmbeddedObject.

Return Value
notesEmbeddedObject   The newly attached file, embedded object, or linked object.

 NOTE
Files can be attached on any Notes platform, but objects and links can only be created on platforms supporting OLE. Files can be embedded as OLE/2 objects only on platforms supporting OLE, but they can be embedded as OLE/1 objects on any Notes platform (if the file is of a supported type, such as a Lotus® application data file).

Creating Embedded Objects Examples
Creating a File Attachment

```
Sub Initialize
'Declarations
    Dim Session As NotesSession
    Dim Db As NotesDatabase
    Dim Doc As NotesDocument
    Dim rtItem As NotesRichTextItem
    Dim EmbeddedObject As NotesEmbeddedObject
'Set Objects
    Set Session = New NotesSession
    Set Db = Session.CurrentDatabase
    Set Doc = New NotesDocument(Db)
'Add Items to document
    Doc.Form = "Memo"
    Doc.Subject = "Rich text example"
'Create rich text item
    Set rtItem = New NotesRichTextItem(Doc,"Body")
    Call rtItem.AppendText("Please read the attached document
to understand our new HR policies.")
    Call rtItem.AddNewLine(2)
'Create a file attachment for the memo
    Set EmbeddedObject = rtItem.EmbObject( EMBED_ATTACHMENT,
"", "c:\hrpolicies.doc")
'Send the document
    Call Doc.Send(False, "Dave Matzl")
End Sub
```

Creating an Embedded Object

```
Sub Initialize
'Declarations
     Dim Session As NotesSession
     Dim Db As NotesDatabase
     Dim Doc As NotesDocument
     Dim rtItem As NotesRichTextItem
     Dim EmbeddedObject As NotesEmbeddedObject
'Set Objects
     Set Session = New NotesSession
     Set Db = Session.CurrentDatabase
     Set Doc = New NotesDocument(Db)
'Add Items to document
     Doc.Form = "Memo"
     Doc.Subject = "Rich text example"
'Create rich text item
     Set rtItem = New NotesRichTextItem(Doc,"Body")
     Call rtItem.AppendText("Please read the attached document
to understand our new HR policies.")
     Call rtItem.AddNewLine(2)
'Embed a word document object
     Set EmbeddedObject = rtItem.EmbedObject( EMBED_OBJECT, "",
"c:\hrpolicies.doc")
'Send the document
     Call Doc.Send(False, "Dave Matzl")
End Sub
```

Creating an Embedded Object Link

```
Sub Initialize
'Declarations
     Dim Session As NotesSession
     Dim Db As NotesDatabase
     Dim Doc As NotesDocument
     Dim rtItem As NotesRichTextItem
     Dim EmbeddedObject As NotesEmbeddedObject
'Set Objects
     Set Session = New NotesSession
     Set Db = Session.CurrentDatabase
     Set Doc = New NotesDocument(Db)
'Add Items to document
     Doc.Form = "Memo"
     Doc.Subject = "Rich text example"
'Create rich text item
     Set rtItem = New NotesRichTextItem(Doc,"Body")
     Call rtItem.AppendText("Please read the attached document
to understand our new HR policies.")
     Call rtItem.AddNewLine(2)
'Embed a word document object link
     Set EmbeddedObject = rtItem.EmbedObject( EMBED_OBJECTLINK,
"", "c:\hrpolicies.doc")
'Send the document
     Call Doc.Send(False, "Dave Matzl")
End Sub
```

Sample Questions

1. *Objective:* Processing document items

   The NotesRichTextItem class inherits from NotesItem, meaning that

   a. NotesRichTextItem objects can use all the values of a declared NotesItem object.

   b. NotesItem objects can use all the properties and methods of NotesRichTextItem.

   c. NotesItem objects all contain a NotesRichTextItem object.

   d. NotesRichTextItem objects can use all the properties and methods of NotesItem.

2. *Objective:* Accessing item values

   John wants to create an item called *Name* on his new NotesDocument object using the extended syntax. If *Doc* is his object reference to a new NotesDocument object, which statement is correct?

   a. Set Doc.Name = "Joe"

   b. Doc.Name(0) = "Joe"

   c. Doc.Name = "Joe"

   d. Doc.Name."Joe"

3. *Objective:* Accessing item values

   Bob is trying to retrieve a value from a field called *LawFirm*. What is wrong with this statement? lawfirmValue = Doc. GetItemValue(LawFirm)

   a. The GetItemValue method is used on the NotesItem class.

   b. There is no method called GetItemValue.

   c. The item name being accessed should be a string and should be enclosed in quotations.

   d. There is nothing wrong with this statement.

4. *Objective:* Accessing item values

Shirley is trying to process a multi-value field called *Judges* using this script:

```
Declarations . . .
MvFieldValues = Doc.Judges(0)-Line 1
Forall j in MvFieldValues-Line 2
    Set newDoc = New NotesDocument(Db)
    NewDoc.JudgeName = j-Line 3
    Call NewDoc.Save(True,False)
End Forall
```

Which line is giving her trouble?

a. Line 1

b. Line 2

c. Line 3

d. The script is correct.

5. *Objective:* Creating and setting item values

The difference between AppendItemValue and ReplaceItemValue is that

a. AppendItemValue appends a value to an existing item, whereas ReplaceItemValue replaces it.

b. Both AppendItemValue and ReplaceItemValue replace an existing item with a new item and value.

c. AppendItemValue creates another item of the same name if there are other items on the document with that name, whereas ReplaceItemValue does not.

d. There is none. They do the same thing.

6. *Objective:* Chris is trying to create a new NotesItem called *SchoolName* of Authors data type. Why is his script not working?

```
Set schoolItem = New NotesItem( Doc, "SchoolName", AUTHORS)
```

a. The first parameter should be the item name, and the second parameter should be the document object.

b. He is missing the value parameter.

c. The AUTHORS data type parameter should be a string value.

d. The script is correct.

7. *Objective:* Rich text items

   Once you set an item to be rich text, you must

   a. Use the NotesItem methods to append text
   b. Use the NotesRichTextItem class methods to append text. You cannot use NotesItem methods to do so
   c. Always use the NotesDocument class methods and properties to access its values
   d. Create a corresponding NotesItem to inherit its values

8. *Objective:* Rich text items

   True or False: All operations on a NotesRichTextItem are immediately visible.

9. *Objective:* Creating rich text items

   Mark is trying to create a rich text item on his *MailMemo* NotesDocument object. What is wrong with his script?

   ```
   Dim RichTextItem As NotesRichTextItem
   Set MailMemo.RichTextItem = New
   NotesRichTextItem(MailMemo,"Body")
   ```

   a. The statement after the equals sign (=) should read *True*, because the RichTextItem is a property of the NotesItem class.
   b. You must set the RichTextItem object and not treat the RichTextItem object as a property of the MailMemo NotesDocument object.
   c. Using *Set* is not acceptable syntax to set a NotesRichTextItem object.
   d. This script is correct.

10. *Objective:* Creating embedded objects

    The NotesEmbeddedObject class represents any one of the following:

    a. An embedded object
    b. An object link
    c. A file attachment
    d. All of the above

Sample Answers

1. *Answer:* c

   The NotesRichTextItem class inherits from NotesItem, meaning that NotesRichTextItem objects can use all the properties and methods of NotesItem.

2. *Answer:* c

   The extended syntax is similar to setting a property for an object. To set a value on a document, use the following syntax: notesDocument.FieldName = value.

3. *Answer:* c

   valueArray = notesDocument.GetItemValue(itemName$)

4. *Answer:* a

   Line 1 is giving her trouble, because when you retrieve values from a multi-value field, you must retrieve them into a variant and then use a looping structure to process them. If the *Judges* field was not a multi-value field, using the zero subscript would be acceptable.

5. *Answer:* c

   AppendItemValue creates another item of the same name, even if there are other items on the document with that name. ReplaceItemValue will replace all items of the same name with the value you specify.

6. *Answer:* c

   You must assign the new NotesItem a value by using the value parameter. The syntax is

   ```
   Set notesItem = New NotesItem( notesDocument, name$, value  [,
   specialType% ] )
   ```

7. *Answer:* b

   Once you set an item to be rich text, you must use the Notes-RichTextItem class methods to append text. You cannot use NotesItem methods to do so. The other three statements are wrong.

8. *Answer:* False

   Operating on rich text fields is not as *visible* as working with non-rich text fields. You generally cannot see the effects of changes made by LotusScript until you save and open the document.

9. *Answer:* d

   You must set the RichTextItem object and not treat the RichText-Item object as a property of the MailMemo NotesDocument object. The correct syntax for the statement is:

   ```
   Set RichTextItem = New NotesRichTextItem(MailMemo,"Body").
   ```

10. *Answer:* b

    The NotesEmbeddedObject class represents any one of the following:

    - An embedded object
    - An object link
    - A file attachment

CHAPTER 8

Controlling Database Access

This chapter reviews controlling database access through the manipulation of the ACL. You can manipulate the ACL of a database by using the NotesACL and NotesACLEntry classes. This chapter covers those classes and some of their properties and methods.

Chapter Objectives

The objectives of this chapter are to increase your understanding of the following items

- Accessing an ACL
- Creating ACL entries
- Changing ACL entries
- Deleting ACL entries
- Adding roles to the ACL
- Removing roles from the ACL
- Associating ACL entries with roles in the ACL

Accessing an ACL

Every Notes database contains a NotesACL object that represents that database's ACL. To get it, you must use the ACL property in the NotesDatabase class.

The NotesDatabase class has three methods you can use to access and modify an ACL without declaring a NotesACL object. All you need to know is the name of the person, server, or group.

- To see what access level a person, server, or group has, use QueryAccess.
- To grant access, use GrantAccess.
- To remove access, use RevokeAccess.

All scripts that are executed against any database ACL assume these access levels of the script owner (if running on a server) or the user executing the script:

- Reader access to query the ACL
- Manager access to change the ACL

You can use various methods to verify the ACL prior to attempting a read or change:

- Attempting to open a NotesDatabase object if the agent/user has been given No Access (use an error-trapping routine to capture this)
- Using the NotesDatabase object CurrentAccessLevel property to determine the level

The Default access level cannot be added, renamed, or deleted. You can perform many other operations on it, however, using the name *Default* with these conditions:

- Use the NotesACLEntry object Level property to change the *Default* setting.
- Do not try to add *Default* using the NotesDatabase object GrantAccess method or the NotesACLEntry New object method.
- Do not try to delete *Default* using the NotesDatabase object RevokeAccess method or the NotesACLEntry object Remove method.

NotesDatabase ACL Property Example

```
Sub Initialize
'Declarations
Dim Db As New NotesDatabase( "Maputo", "notefile\none.nsf" )
Dim ACL As NotesACL
Dim Entry As NotesACLEntry
'Get a handle on the database's ACL
Set ACL = Db.ACL
'Get the first entry in the ACL
Set Entry = ACL.GetFirstEntry
End Sub
```

Finding an ACL Entry

Finding the access level of a specified entry, whether in the current database or in another database, can return the highest resolved entry or the specific entry level.

If you want to find the highest resolved entry, use the NotesDatabase class QueryAccess method.

NotesDatabase Class QueryAccess Method

Returns a person's, group's, or server's current access level to a database.

Syntax

```
level% = notesDatabase.QueryAccess( name$ )
```

Parameters

name$ **String.**   The name of the person, group, or server.

Return Values

level%   Integer constant. Indicates the current access level and is one of the following items, as seen in Table 8.1.

Table 8.1 ACL Level Constants

ACLLEVEL_NOACCESS	**No access**
ACLLEVEL_DEPOSITOR	**Depositor access**
ACLLEVEL_READER	**Reader access**

continues

Table 8.1 Continued

`ACLLEVEL_AUTHOR`	**Author access**
`ACLLEVEL_EDITOR`	**Editor access**
`ACLLEVEL_DESIGNER`	**Designer access**
`ACLLEVEL_MANAGER`	**Manager access**

The NotesDatabase class QueryAccess method will always return an access level. You must match character-for-character the exact entry (including case) when using the NotesACL class GetEntry method to return a level. The highest resolved ACL level is the result of this algorithm:

- Membership in any group takes precedence over the *Default* setting, whether higher or lower.
- Users who are listed in two or more groups in the ACL are granted the higher access level.
- Users specifically listed in the ACL are granted the specific rights assigned (takes precedence over any group memberships).

Finding an ACL Entry Example

```
Sub Initialize
      'Anticipate entry not in ACL which results in object
variable not set
      On Error Resume Next
'Have user enter a user name, group or server
      GetUser = Inputbox("Enter the name of a user, group, or
server:", "Get Entry","")
'Declarations
      Dim Session As NotesSession
      Dim Database As NotesDatabase
      Dim ACL As NotesACL
      Dim Entry As NotesACLEntry
'Set objects
      Set Session = New NotesSession
      Set Database = Session.CurrentDatabase
      Set ACL = Database.ACL
'Gets the highest level of access for a user
      GetHighestLevel = Database.QueryAccess(GetUser)
'Format the level in a more understandable format using the
Case statement
      Select Case GetHighestLevel
      Case ACLLEVEL_NOACCESS
```

```
            ReturnLevel = "No Access"
    Case ACLLEVEL_DEPOSITOR
            ReturnLevel = "Depositor"
    Case ACLLEVEL_READER
            ReturnLevel =  "Reader"
    Case ACLLEVEL_AUTHOR
            ReturnLevel = "Author"
    Case ACLLEVEL_EDITOR
            ReturnLevel = "Editor"
    Case ACLLEVEL_DESIGNER
            ReturnLevel = "Designer"
    Case ACLLEVEL_MANAGER
            ReturnLevel = "Manager"
    Case Else
            ReturnLevel = "Unknown"
    End Select
'Prompt the user with the access level
    Messagebox GetUser & "access level is " & GetHighestLevel
& " access."
End Sub
```

This example takes input from the user and checks the input against the ACL by using the QueryAccess method. The process then uses a case statement to return the access level in a more meaningful manner that can be displayed to the user in a message box.

Reading the ACL

If you do not know the entries in an ACL, you can recursively access the entries in the same way you access a data directory or view. Start with the first item in the object, and then use the *Get-Next . . .* method.

Recursive ACL Access Example

```
Sub Click(Source As Button)
'Declarations
    Dim Session As New NotesSession
    Dim Database As NotesDatabase
    Dim  ACL As NotesACL
    Dim Entry As NotesACLEntry
'Set objects
    Set Database = Session.CurrentDatabase
    Set ACL = Database.ACL
    Set Entry = ACL.GetFirstEntry
'Recursively access the ACL and display the access level
    Do Until Entry Is Nothing
            GetLevel% = Entry.Level
            Messagebox Entry.Name & ":  " &
LabelACLLevel(GetLevel%)
```

```
            Set Entry = ACL.GetNextEntry(Entry)
        Loop
    End Sub

    Function LabelACLLEVEL (Level As Integer)
        Select Case Level
        Case 0
            LabelACLLevel = "No Access"
        Case 1
            LabelACLLevel = "Depositor"
        Case 2
            LabelACLLevel = "Reader"
        Case 3
            LabelACLLevel = "Author"
        Case 4
            LabelACLLevel = "Editor"
        Case 5
            LabelACLLevel = "Designer"
        Case 6
            LabelACLLevel = "Manager"
        End Select
    End Function
```

In this example, the script recursively accesses a database's ACL using the GetNextEntry method—and prompts the user with the ACL level of the users in the ACL. Notice that a function called LabelACLLevel is used to turn the numeric return value of the Level property into a more meaningful value, so that it can be displayed to the user.

Adding, Changing, and Removing ACL Entries

You can use either the NotesDatabase class GrantAccess method or the NotesACLEntry New object method to add a new entry to an ACL.

Grant Access Method

Syntax
```
    Call notesDatabase.GrantAccess( name$, level% )
```

Parameters
name$ **String.**   The name of the person, group, or server whose access level you want to provide or change.

level%   Integer constant. The level of access you are granting. This parameter is one of the following constants, as shown in Table 8.2.

Table 8.2 ACL Level Constants

ACLLEVEL_NOACCESS	**No access**
ACLLEVEL_DEPOSITOR	**Depositor access**
ACLLEVEL_READER	**Reader access**
ACLLEVEL_AUTHOR	**Author access**
ACLLEVEL_EDITOR	**Editor access**
ACLLEVEL_DESIGNER	**Designer access**

New NotesACLEntry

To create a new NotesACLEntry object, use one of the following commands

- New
- CreateACLEntry method in NotesACL

New creates an entry in an ACL with the name and level that you specify. You must call Save on the ACL if you want the modified ACL to be saved to disk.

Syntax

```
Dim variableName As New NotesACLEntry( notesACL, name$, level% )
```

or

```
Set notesACLEntry = New NotesACLEntry( notesACL, name$, level% )
```

Parameters

notesACL   The ACL that contains the entry you are creating.

name$ **String.**   The name of the person, group, or server for whom you want to create an entry in the ACL.

level% Constant. The level you want to assign to this person, group, or server in the ACL. May be any of the following items, as shown in Table 8.3.

Table 8.3 ACL Level Constants

`ACLLEVEL_NOACCESS`	**No access**
`ACLLEVEL_DEPOSITOR`	**Depositor access**
`ACLLEVEL_READER`	**Reader access**
`ACLLEVEL_AUTHOR`	**Author access**
`ACLLEVEL_EDITOR`	**Editor access**
`ACLLEVEL_DESIGNER`	**Designer access**
`ACLLEVEL_MANAGER`	**Manager access**

GrantAccess Example

```
Sub Click(Source As Button)
'Declarations
    Dim Session As NotesSession
    Dim Database As NotesDatabase
    Dim Level As Integer
'Set objects
    Set Session = New NotesSession
    Set Database = Session.CurrentDatabase
'Get an ACL entry name from the user
    GetUser = Inputbox("Enter the name of a user, group, or
server:", "Grant Entry","")
'Get the access level the user should have from the user
'Note that this loop will not stop until a user enters a value
that is contained in the Instr() function
    StopLoop = 0
    Do Until StopLoop=1
        GetLevel = Inputbox("Enter the level you want the
entry to have:", "Get Level","")
        If Instr("No
AccessDepositorReaderAuthorEditorDesignerManager", GetLevel)
Then
                StopLoop=1
        End If
    Loop
'Reformat the access level the user entered into a constant
    Select Case GetLevel
    Case "No Access"
        Level=ACLLEVEL_NOACCESS
    Case "Depositor"
        Level=ACLLEVEL_DEPOSITOR
    Case "Reader"
        Level=ACLLEVEL_READER
    Case "Author"
        Level=ACLLEVEL_AUTHOR
    Case "Editor"
```

```
            Level=ACLLEVEL_EDITOR
     Case "Designer"
            Level=ACLLEVEL_DESIGNER
     Case "Manager"
            Level=ACLLEVEL_MANAGER
     End Select
     If GetUser="" And GetLevel="" Then
            Exit Sub
     Else
'Grant the access and prompt the user
            Call Database.GrantAccess (GetUser, Level)
            Print GetUser & " was given " & GetLevel & " access."
     End If
End Sub
```

This example uses the GrantAccess method to add a user to the ACL. Notice that the Case statement takes the user's access-level entry and sets the Level variable to an ACL-level constant.

New NotesACLEntry Example

```
'Declarations
     Dim Session As NotesSession
     Dim Database As NotesDatabase
     Dim ACL As NotesACL
     Dim NewEntry As NotesACLEntry
     Dim Level As Integer
'Set objects
     Set Session = New NotesSession
     Set Database = Session.CurrentDatabase
     Set ACL = Database.ACL
'Get an ACL entry name from the user
     GetUser = Inputbox("Enter the name of a user, group, or
server:", "Grant Entry","")
'Get the access level the user should have from the user
'Note that this loop will not stop until a user enters a value
that is contained in the Instr() function
     StopLoop = 0
     Do Until StopLoop=1
            GetLevel = Inputbox("Enter the level you want the
entry to have:", "Get Level","")
            If Instr("No
AccessDepositorReaderAuthorEditorDesignerManager", GetLevel)
Then
                StopLoop=1
            End If
     Loop
'Reformat the access level the user entered into a constant
     Select Case GetLevel
     Case "No Access"
            Level=ACLLEVEL_NOACCESS
     Case "Depositor"
            Level=ACLLEVEL_DEPOSITOR
     Case "Reader"
            Level=ACLLEVEL_READER
```

```
        Case "Author"
                Level=ACLLEVEL_AUTHOR
        Case "Editor"
                Level=ACLLEVEL_EDITOR
        Case "Designer"
                Level=ACLLEVEL_DESIGNER
        Case "Manager"
                Level=ACLLEVEL_MANAGER
        End Select
        If GetUser="" And GetLevel="" Then
                Exit Sub
        Else
'Grant the access and prompt the user
                Set NewEntry = New NotesACLEntry(ACL, GetUser, Level)
                Call ACL.Save
                Print GetUser & " was given " & GetLevel & " access."
        End If
End Sub
```

This script example uses the New NotesACLEntry method to add a user to the ACL.

Changing an ACL Entry

You can change the level on an existing entry in two ways:

- The NotesDatabase class GrantAccess method (an existing entry is given to the new level, and if the entry does not exist, it is added).

- NotesACLEntry Class Level property

GrantAccess Example

```
Sub Click(Source As Button)
'Declarations
        Dim Session As NotesSession
        Dim Database As NotesDatabase
        Dim Level As Integer
'Set objects
        Set Session = New NotesSession
        Set Database = Session.CurrentDatabase
'Get an ACL entry name from the user
        GetUser = Inputbox("Enter the name of a user, group, or
server:", "Grant Entry","")
'Get the access level the user should have from the user
'Note that this loop will not stop until a user enters a value
that is contained in the Instr() function
        StopLoop = 0
        Do Until StopLoop=1
                GetLevel = Inputbox("Enter the level you want the
entry to have:", "Get Level","")
                If Instr("No
```

```
AccessDepositorReaderAuthorEditorDesignerManager", GetLevel)
Then
                    StopLoop=1
          End If
     Loop
'Reformat the access level the user entered into a constant
     Select Case GetLevel
     Case "No Access"
          Level=ACLLEVEL_NOACCESS
     Case "Depositor"
          Level=ACLLEVEL_DEPOSITOR
     Case "Reader"
          Level=ACLLEVEL_READER
     Case "Author"
          Level=ACLLEVEL_AUTHOR
     Case "Editor"
          Level=ACLLEVEL_EDITOR
     Case "Designer"
          Level=ACLLEVEL_DESIGNER
     Case "Manager"
          Level=ACLLEVEL_MANAGER
     End Select
     If GetUser="" And GetLevel="" Then
          Exit Sub
     Else
'Grant the access and prompt the user
          Call MyDatabase.GrantAccess (GetUser, Level)
          Print GetUser & " was given " & GetLevel & " access."
     End If
End Sub
```

This example uses the GrantAccess method to change a user's access in the ACL. If the user does not exist, the GrantAccess method will add the user.

New NotesACLEntry Example

```
'Declarations
     Dim Session As NotesSession
     Dim Database As NotesDatabase
     Dim ACL As NotesACL
     Dim Entry As NotesACLEntry
     Dim Level As Integer
'Set objects
     Set Session = New NotesSession
     Set Database = Session.CurrentDatabase
     Set ACL = Database.ACL
'Get an ACL entry name from the user
     GetUser = Inputbox("Enter the name of a user, group, or
server:", "Grant Entry","")
'Get the access level the user should have from the user
'Note that this loop will not stop until a user enters a value
that is contained in the Instr() function
     StopLoop = 0
     Do Until StopLoop=1
          GetLevel = Inputbox("Enter the level you want the
```

```
entry to have:", "Get Level","")
          If Instr("No
AccessDepositorReaderAuthorEditorDesignerManager", GetLevel)
Then
                StopLoop=1
          End If
     Loop
'Reformat the access level the user entered into a constant
     Select Case GetLevel
     Case "No Access"
          Level=ACLLEVEL_NOACCESS
     Case "Depositor"
          Level=ACLLEVEL_DEPOSITOR
     Case "Reader"
          Level=ACLLEVEL_READER
     Case "Author"
          Level=ACLLEVEL_AUTHOR
     Case "Editor"
          Level=ACLLEVEL_EDITOR
     Case "Designer"
          Level=ACLLEVEL_DESIGNER
     Case "Manager"
          Level=ACLLEVEL_MANAGER
     End Select
     If GetUser="" And GetLevel="" Then
          Exit Sub
     Else
'Change the access and prompt the user
Set Entry = ACL.GetUser(GetUser)
Entry.Level = Level
Call ACL.Save
Print GetUser & " was given " & GetLevel & " access."
     End If
End Sub
```

This script example uses the NotesACLEntry method to change a user's access in the ACL.

Setting Access Flags

The NotesACLEntry class has the following properties that enable you to change the user access flags:

- CanCreateDocuments
- CanCreatePersonalAgent
- CanCreatePersonalFolder
- CanDeleteDocuments
- IsPublicReader
- IsPublicWriter

All of these flags are set using the Boolean True/False values.

Syntax

To get: flag = notesACLEntry.Property

To set: notesACLEntry.Property = flag

Setting Access Flags Example

```
Sub Initialize
'Anticipate entry not in ACL which results in object variable
not set
     On Error Resume Next
'Have user enter a user name, group or server
     GetUser = Inputbox("Enter the name of a user, group, or
server:", "Get Entry","")
'Declarations
     Dim Session As NotesSession
     Dim Database As NotesDatabase
     Dim ACL As NotesACL
     Dim Entry As NotesACLEntry
'Set objects
     Set Session = New NotesSession
     Set Database = Session.CurrentDatabase
     Set ACL = Database.ACL
'Get entry in ACL
     Set Entry = ACL.GetEntry(GetUser)
'Set Access Flags
     Entry.CanCreateDocuments = True
     Entry.CanCreatePersonalAgent = False
     Entry.CanCreatePersonalFolder = True
     Entry.CanDeleteDocuments = False
     Entry. IsPublicReader = True
     Entry.IsPublicWriter = False
'Save the ACL
     Call ACL.Save
End Sub
```

Removing an ACL Entry

To remove an entry from an ACL, you can use

- NotesDatabase class RevokeAccess method
- NotesACLEntry class Remove method

NotesDatabase Class RevokeAccess Example

```
Sub Initialize
'Have user enter a user name, group or server
     GetUser = Inputbox("Enter the name of a user, group, or
server:", "Get Entry","")
'Declarations
     Dim Session As NotesSession
```

```
    Dim Database As NotesDatabase
    Dim ACL As NotesACL
    Dim Entry As NotesACLEntry
'Set objects
    Set Session = New NotesSession
    Set Database = Session.CurrentDatabase
'Revoke Access
    Call Database.RevokeAccess(GetUser)
End Sub
NotesACLEntry class Remove Example
Sub Initialize
'Have user enter a user name, group or server
    GetUser = Inputbox("Enter the name of a user, group, or
server:", "Get Entry","")
'Declarations
    Dim Session As NotesSession
    Dim Database As NotesDatabase
    Dim ACL As NotesACL
    Dim Entry As NotesACLEntry
'Set objects
    Set Session = New NotesSession
    Set Database = Session.CurrentDatabase
    Set ACL = Database.ACL
'Get entry in ACL
    Set Entry = ACL.GetEntry(GetUser)
'Remove the user from the ACL
    Call Entry.Remove(GetUser)
'Save the ACL
    Call ACL.Save
End Sub
```

Roles in the ACL

Adding, Removing, and Renaming a Role in an ACL

To add, remove, or rename a role to the ACL in a database, you need to use the NotesACL class:

- AddRole method
- DeleteRole method
- RenameRole method

AddRole Method
Syntax
```
    Call notesACL.AddRole( name$ )
```

Parameter
name$ **String.**   The name of the new role. Do not put parentheses or square brackets around the name.

DeleteRole Method
Syntax
```
Call notesACL.DeleteRole( name$ )
```

Parameter
name$ **String.** The name of the role to remove.

If the role you specified does not exist in the ACL, Notes raises the error, *Role name not found.*

RenameRole Method
Syntax
```
Call notesACL.RenameRole( oldName$, newName$ )
```

Parameters
oldName$ **String.** The current name of the role.

newName$ **String.** The new name you want to give to the role.

When you rename a role, any entries in the ACL that had the old role will get the new role.

Add, Remove, and Rename Role Examples

Add Role in ACL Example
```
Sub Initialize
'Declarations
    Dim Session As New NotesSession
    Dim Database As NotesDatabase
    Dim ACL As NotesACL
'Set objects
    Set Database = Session.CurrentDatabase
    Set ACL = MyDatabase.ACL
'Get name of role from user
    GetRole$ = Inputbox("Enter the name of the role you wish
to add","Add Role")
'Add Role to ACL and save it
    Call ACL.AddRole(GetRole$)
    Call ACL.Save
'Prompt user of change
    Print "Role " & GetRole$ & " added to " & Database.Title &
" ."
End Sub
```

Remove Role in ACL Example
```
Sub Initialize
'Declarations
    Dim Session As New NotesSession
    Dim Database As NotesDatabase
    Dim ACL As NotesACL
'Set objects
    Set Database = Session.CurrentDatabase
```

```
    Set ACL = MyDatabase.ACL
'Get name of role from user
    GetRole$ = Inputbox("Enter the name of the role you wish
to delete","Add Role")
'Add Role to ACL and save it
    Call ACL.DeleteRole(GetRole$)
    Call ACL.Save
'Prompt user of change
    Print "Role " & GetRole$ & " deleted from " &
Database.Title & " ."
End Sub
```

Rename Role in ACL Example

```
Sub Initialize
'Declarations
    Dim Session As New NotesSession
    Dim Database As NotesDatabase
    Dim ACL As NotesACL
'Set objects
    Set Database = Session.CurrentDatabase
    Set ACL = MyDatabase.ACL
'Get name of role from user
    GetRole$ = Inputbox("Enter the name of the role you wish
to change","Rename Role")
    NewName$ = Inputbox("Enter the name you would like to
change " & GetRole$ & " to","Rename Role")
'Add Role to ACL and save it
    Call ACL.RenameRole(GetRole$,NewName$)
    Call ACL.Save
'Prompt user of change
    Print "Role " & GetRole$ & " rename to " & NewName$ & " ."
End Sub
```

Associating and Disassociating a User with a Role in an ACL

To associate or disassociate a user with a role in an ACL, use the NotesACLEntry class

- EnableRole method
- DisableRole method

EnableRole Method
Syntax
```
    Call notesACLEntry.EnableRole( name$ )
```

Parameter
name$ **String.**   The name of the role to enable.

If the role does not exist in the ACL, this method raises the error, *Role name not found.*

If the role exists in the ACL and is already enabled for the entry, this method does nothing.

DisableRole Method
Syntax
```
Call notesACLEntry.DisableRole( name$ )
```

Parameter
name$ **String.**   The name of the role to disable.

If the role does not exist in the ACL, this method raises the error, *Role name not found.*

If the role exists in the ACL but is already disabled for the entry, this method does nothing.

Role Association Examples
EnableRole Example
```
Sub Initialize
'Have user enter a user name, group or server
    GetUser = Inputbox("Enter the name of a user, group, or
server:", "Get Entry","")
'Have user enter a role name
    Role = Inputbox("Enter the name of the role you want to
add " & GetUser & " to","Add to Role")
'Declarations
    Dim Session As NotesSession
    Dim Database As NotesDatabase
    Dim ACL As NotesACL
    Dim Entry As NotesACLEntry
'Set objects
    Set Session = New NotesSession
    Set Database = Session.CurrentDatabase
    Set ACL = Database.ACL
'Get entry in ACL
    Set Entry = ACL.GetEntry(GetUser)
'Add GetUser to role
    Call Entry.EnableRole(Role)
'Save the ACL
    Call ACL.Save
End Sub
```

DisableRole Example
```
Sub Initialize
'Have user enter a user name, group or server
    GetUser = Inputbox("Enter the name of a user, group, or
server:", "Get Entry","")
'Have use enter a role name
    Role = Inputbox("Enter the name of the role you want to
```

```
remove " & GetUser & " from","Remove")
'Declarations
     Dim Session As NotesSession
     Dim Database As NotesDatabase
     Dim ACL As NotesACL
     Dim Entry As NotesACLEntry
'Set objects
     Set Session = New NotesSession
     Set Database = Session.CurrentDatabase
     Set ACL = Database.ACL
'Get entry in ACL
     Set Entry = ACL.GetEntry(GetUser)
'Add GetUser to role
     Call Entry.DisableRole(Role)
'Save the ACL
     Call ACL.Save
End Sub
```

Sample Questions

1. *Objective:* Accessing an ACL

   The NotesDatabase class has three methods you can use to modify an ACL without declaring a NotesACL object. The method that is not one of these is

   a. EnableRole

   b. RevokeAccess

   c. GrantAccess

   d. QueryAccess

2. *Objective:* Accessing an ACL

   All scripts that are executed against any database ACL assume these two access levels:

   a. Reader

   b. Author

   c. Designer

   d. Manager

3. *Objective:* Accessing an ACL

   The Default access level

   a. Must be added

   b. Can be deleted

   c. Cannot be added

   d. Can be renamed to something more meaningful

4. *Objective:* Finding an ACL entry

   The NotesDatabase QueryAccess method

   a. Returns an entry's access level

   b. Returns the highest resolved access for an entry

   c. Returns all of the Roles used by an entry

   d. Returns the access flags for an entry

5. *Objective:* Adding an ACL entry

   The correct syntax for the GrantAccess method is

   a. `Call notesACLEntry.GrantAccess( name$, level% )`

   b. `Call notesDatabase.GrantAccess( name% )`

   c. `Call notesACLEntry.GrantAccess( level% )`

   d. `Call notesDatabase.GrantAccess( name$, level% )`

6. *Objective:* Notes ACL entry

   To create a new NotesACLEntry object, use one of the following items

   a. CreateACLEntry method in NotesDatabase class

   b. CreateACLEntry method in NotesACLEntry class

   c. CreateACLEntry method in NotesACL class

   d. New method in the NotesACL class

7. *Objective:* Access flags

   The _____ class has properties that enable you to change the user access flags.

   a. NotesACL

   b. NotesACLEntry

   c. NotesDatabase

   d. All of the above

8. *Objective:* Removing an ACL entry.

   True or False: To remove an entry from the ACL, you can use the NotesDatabase class RevokeAccess method or the NotesACLEntry class Remove method.

9. *Objective:* Roles

   The NotesACL class has a method for

   a. Adding a role
   b. Removing a role
   c. Renaming a role
   d. All of the above

10. *Objective:* Associating roles with users

   To associate a user with a role in an ACL, use the NotesACLEntry class

   a. AddToRole method
   b. PutInRole method
   c. EnableRole method
   d. None of the above

Sample Question Answers

1. *Answer:* a

   The NotesDatabase class has three methods you can use to access and modify an ACL without declaring a NotesACL object. All you need to know is the name of the person, server, or group.

   - To see what access level a person, server, or group has, use QueryAccess.
   - To grant access, use GrantAccess.
   - To remove access, use RevokeAccess.

2. *Answers:* a and d

   All scripts that are executed against any database ACL assume these access levels of the script owner (if running on a server) or the user executing the script

   - Reader access to query the ACL
   - Manager access to change the ACL

3. *Answer:* c

   The Default access level cannot be added, renamed, or deleted. You can perform many other operations on it, however, using the name *Default* with these conditions:

   - Use the NotesACLEntry object Level property to change the *Default* setting.
   - Do not try to add *Default* using the NotesDatabase object GrantAccess method or the NotesACLEntry New object method.
   - Do not try to delete *Default* by using the NotesDatabase object RevokeAccess method or the NotesACLEntry object Remove method.

4. *Answer:* b

   Returns a person's, group's, or server's current access level to a database (the highest resolved access level)

5. *Answer:* d

```
Call notesDatabase.GrantAccess( name$, level% ).
```

6. *Answer:* c

To create a new NotesACLEntry object, use one of the following commands:

- New
- CreateACLEntry method in NotesACL

New creates an entry in an ACL with the name and level that you specify. You must call Save on the ACL if you want the modified ACL to be saved to disk.

7. *Answer:* b

The NotesACLEntry class has the following properties that enable you to change the user access flags

- CanCreateDocuments
- CanCreatePersonalAgent
- CanCreatePersonalFolder
- CanDeleteDocuments
- IsPublicReader
- IsPublicWriter

8. *Answer:* True

To remove an entry from an ACL, you can use the following methods:

- NotesDatabase class RevokeAccess method
- NotesACLEntry class Remove method

9. *Answer:* d

   To add, remove, or rename a role to the ACL in a database, you need to use the NotesACL class

   - AddRole method
   - DeleteRole method
   - RenameRole method

10. *Answer:* c

    To associate or disassociate a user with a role in an ACL, use the NotesACLEntry class

    - EnableRole method
    - DisableRole method

CHAPTER 9

Session Access

This chapter reviews session access. The NotesSession class provides a means for accessing attributes of the environment and persistent information about agents, besides providing methods for accessing environment variables.

Chapter Objectives

The objectives of this chapter are to increase your understanding of the following items:

- Accessing session properties
- Writing to the environment
- Retrieving values from the environment

Session Information

The session is the environment in which the current script is running. Using the NotesSession class properties, you can test for various conditions and set environment variables.

NotesSession Creation and Access

To access the current session, use New.

Syntax

```
Dim variableName As New NotesSession
```

or

```
Set notesSession = New NotesSession
```

To access the current session from a NotesDatabase object, use the Parent property in the NotesDatabase.

Usage

Because there can be only one session per script, the New method will always return the same object each time you call it.

 NOTE
You do not need the session handle to access databases or perform other operations in Notes—only to access session properties and methods.

User Environment Control

The user environment is either

- The server or workstation containing the script's database in the following cases: an agent whose trigger is If New Mail Has Arrived, or an agent whose trigger is On Schedule

- The workstation of the current user in all other cases

Using Environment Variables

The GetEnvironmentValue, GetEnvironmentString, and SetEnvironmentVar methods retrieve and set environment variables, which are stored in the local NOTES.INI or Preferences file.

 NOTE
Use GetEnvironmentValue only for numeric environment variables. Use GetEnvironmentString for strings and numeric values.

Environment variables are useful for saving data between Notes sessions on a single server or workstation, where no conflicts are pos-

sible. They are also useful for obtaining the environment information set by Notes, such as KitType, Directory, Preferences, Domain,
Port, and others.

 NOTE
For the GetEnvironmentValue, GetEnvironmentString, and Set-
EnvironmentVar methods, if the script runs on a workstation, the
method retrieves the environment variable from the current user's
NOTES.INI (or Notes Preferences) file. If the script runs on a server, the
method retrieves the environment variable from the server's NOTES.INI file,
subject to administrative restrictions (see "Restricting agents on servers" in
Notes Administration Help or the "Administrator's Guide").

GetEnvironmentValue Method
Syntax
```
valueV = notesSession.GetEnvironmentValue( name$ [ . system ] )
```

Parameter

name$ **String.** The name of the environment variable to get.

system **Boolean.** Optional. If True, the method uses the exact
name$ of the environment variable. If False, the method assigns $
to the name$ before retrieving its value. If you omit this parameter,
it defaults to False.

Return Value
valueV Variant. The value of the environment variable.

GetEnvironmentValue Example
```
Sub Initialize
'Declarations
     Dim session As New NotesSession
     Dim EmployeeNumber As Integer
'Retrieve the numeric EmployeeNum environment variable from the
ini file
     EmployeeNumber = session.GetEnvironmentValue
( "EmployeeNum" )
End Sub
```

GetEnvironmentString Method
Syntax
```
valueV = notesSession.GetEnvironmentString( name$ [ . system ] )
```

Parameters
name$ **String.** The name of the environment variable to get.

system **Boolean.** Optional. If True, the method uses the exact
name$ of the environment variable. If False, the method assigns $

to the *name$* before retrieving its value. If you omit this parameter, it defaults to False.

Return Value

valueV   Variant. The value of the environment variable.

GetEnvironmentString Example

```
Sub Initialize
'Declarations
     Dim session As New NotesSession
     Dim location As String
'Retrieve the PlantLocation environment variable value
     location = session.GetEnvironmentString( "PlantLocation" )
End Sub
```

SetEnvironmentVar Method

Syntax

```
Call notesSession.SetEnvironmentVar( name$, valueV
[, issystemvar] )
```

Parameters

name$ **String.**   The name of the environment variable to set.

valueV   Variant. The value of the environment variable. Date values are converted to strings.

issystemvar **Boolean.**   Optional. If True, no dollar sign (*$*) is assigned to the variable name. Defaults to False.

If an environment variable called *name$* already exists, it gets the new value; otherwise, a new environment variable is added to the NOTES.INI (or Notes Preferences) file.

SetEnvironmentVar assigns a dollar sign (*$*) to the *name$* before setting the value of the environment variable, unless you specify *issystemvar* as False or have already included a *$* as the first character of the *name$*. When you examine the environment variable in the NOTES.INI (or Notes Preferences) file, you see the *$* character. When retrieving environment variables with the GetEnvironmentString or GetEnvironmentValue methods, you have the option of including or omitting the *$* from the variable name.

If a script runs on a workstation, the user's environment variable is set. If a script runs on a server, the server's environment variable is set subject to administrative restrictions (see "Restricting Agents on Servers" in Notes Administration Help or the "Administrator's Guide").

TROUBLESHOOTING TIP
If *valueV* is not an appropriate data type, SetEnvironmentVar raises the error: *Environment variables must be strings, dates, or integers.*

SetEnvironmentVar Example

```
Sub Initialize
'Declarations
     Dim session As New NotesSession
'Set the environment variable UserName with the current user's
name
     Call session.SetEnvironmentVar( "UserName",
Session.UserName )
End Sub
```

Current Agent Information

Information about the current agent is available using the Notes-Session object properties CurrentAgent, LastExitStatus, and Lastrun. This information can be used to

- Get a handle on the agent that is currently running
- Determine the exit status code returned by the Agent Manager the last time the current agent ran
- Determine the date when the current agent was last executed

CurrentAgent Property

Read-only. The agent that is currently running.

Syntax

To get: `Set notesAgent = notesSession.CurrentAgent`

CurrentAgent Example

```
Sub Initialize
'Declarations
     Dim session As New NotesSession
     Dim agent As NotesAgent
     Dim agentName As String
'Use the CurrentAgent property to set the agent object
     Set agent = session.CurrentAgent
'Get the agent name
     agentName = agent.Name
End Sub
```

LastExitStatus Property

Read-only. The exit status code returned by the Agent Manager the last time the current agent ran.

Syntax

To get: *code%* = *notesSession*.LastExitStatus

NOTE
This property returns zero if the agent ran without errors.

LastExitStatus Example

```
Sub Initialize
'Declarations
    Dim session As New NotesSession
'Check the LastExitStatus
    If ( session.LastExitStatus = 0 ) Then
        Print ( "Agent exited gracefully last time" )
    Else
        Print ( "Agent raised an error last time" )
    End If
End Sub
```

LastRun Property

Read-only. The date when the current agent was last executed.

Syntax

To get: *dateV* = *notesSession.LastRun*

NOTE
If the script has never been run before, this property returns a variant containing the date-time value 11/30/1899.

LastRun Example

```
Sub Initialize
'Declarations
    Dim session As New NotesSession
    Dim db As NotesDatabase
'Set the db object
    Set db = session.CurrentDatabase
'Check the LastRan property
    If ( session.LastRun > db.LastFTIndexed ) Then
        Call db.UpdateFTIndex( False )
    End If
End Sub
```

Saved Data

Agents often require information about conditions encountered when the agent last ran.

If, for example, the NotesSession object LastExitStatus is False (zero), then the agent ran last time without errors. If there were errors, it might make sense not to run the agent again (especially if the agent consumes a significant amount of server resources).

This information is stored and saved by the agent itself. You may want more detailed information available to an agent when it runs —information not about the agent itself, but about other conditions found in the environment or the data.

You can store this type of information in several places:

- In the environment, using the NotesSession class SetEnvironment-Var/GetEnvironmentString methods

- In any document in any Notes database, using the NotesDocument class methods

- In a text file, using LotusScript commands (for local agents only)

- In the agent itself, using the NotesSession class SavedData property

To use this property, think of the agent itself having the capacity to store any item you choose. You can then retrieve it the next time the agent runs. If the agent is modified, however, all the assignments are cleared.

The SavedData property has the advantage over all other ways to store data, in that it

- Is the most portable method (works on a server or workstation without modification)

- Is easy to code, because you do not have to locate a file or document before accessing the information

- Is reinitialized each time the agent is modified

Syntax
To get: `Set notesDocument = notesSession.SavedData`

 NOTE
This property is valid only for agent scripts. In any other script, SavedData returns nothing.

The SavedData document is created when you save an agent, and it is stored in the same database as the agent. The document replicates but is not displayed in views.

Each time you edit and resave an agent, its SavedData document is deleted—and a new, blank document is created. When you delete an agent, its SavedData document is deleted.

SavedData Example

```
Sub Initialize
'Declarations
    Dim session As New NotesSession
    Dim doc As NotesDocument
'Create the saved data document
    Set doc = session.SavedData
'Add an item to the saved data document to store the date
    doc.RunDate = Date
    Call doc.Save( True, True )
End Sub
```

Sample Questions

1. *Objective:* NotesSession

   True or False: You need the session handle to access databases or perform other operations in Notes.

2. *Objective:* NotesSession

   The user environment is

   a. The workstation of all users
   b. The server of all users
   c. The workstation of the current user
   d. None of the above

3. *Objective:* Using environment variables

   The method for retrieving numeric environment variables is

   a. GetEnvironmentString
   b. GetEnvironmentValue
   c. GetEnvironmentNumber
   d. None of the above

4. *Objective:* NotesSession properties

   Which property returns the status of an agent the last time it ran?

   a. LastRun
   b. LastRunStatus
   c. LastExitStatus
   d. AgentStatus

5. *Objective:* NotesSession properties

   The SavedData property has the advantage over all other ways to store data, in that it

   a. Is the most portable method (works on a server or workstation without modification)
   b. Is easy to code because you do not have to locate a file or document before accessing the information
   c. Is reinitialized each time the agent is modified
   d. All of the above

Sample Answers

1. *Answer:* False

   You do not need the session handle to access databases or perform other operations in Notes—only to access session properties and methods.

2. *Answer:* c

   The user environment is either

   - The server or workstation containing the script's database in the following cases: an agent whose trigger is If New Mail Has Arrived, and an agent whose trigger is On Schedule
   - The workstation of the current user in all other cases

3. *Answer:* b

   Use GetEnvironmentValue only for numeric environment variables. Use GetEnvironmentString for strings and numeric values

4. *Answer:* b

   The LastExitStatus will return the exit status code returned by the Agent Manager the last time the current agent ran.

5. *Answer:* d

   The SavedData property has the advantage over all other ways to store data, in that it

   - Is the most portable method (works on a server or workstation without modification)
   - Is easy to code, because you do not have to locate a file or document before accessing the information
   - Is reinitialized each time the agent is modified

CHAPTER 10

Accessing Non-Notes Data

This chapter reviews the *LotusScript:DataObject* (LS:DO) classes used in LotusScript. The LS:DO provides full read and write access to external *Open Database Connectivity* (ODBC) data sources, using the complete control and flexibility of the structured programming language LotusScript.

TIP
There are questions on the exam that relate to the concepts and coding discussed in this chapter. If you have not used the LS:DO, you should take the time to create connections to non-Notes databases and become familiar with the concepts before taking the exam.

Chapter Objectives

The objectives of this chapter are to increase your understanding of the following items:

- Loading the LSXModule
- Accessing non-Notes data
- Creating an ODBC query
- Returning result sets from non-Notes databases
- Reading values from a result set

LotusScript:DataObject (LS:DO)

The LS:DO consists of a set of three classes: ODBCConnection, ODBCQuery, and ODBCResultSet. These classes provide properties and methods for accessing, editing, updating, and creating tables in external databases through the ODBC v2.0 standard.

> **NOTE**
> You must put the following statement in the (Options) event of the (Global) object to access the OBDC classes: `UseLSX "*LSX-ODBC"`. The leading "*" tells LotusScript to use the class registry to look up the path of the LS:DO dynamic library being loaded. This method is a platform-independent way of loading LS:DO, because each operating system uses different methods.

You can test for and handle all ODBC errors with On Error, the same as for Notes methods. In addition, the ODBC methods return a Boolean status code to indicate success or failure—unless there is another return—and provide the methods GetError, GetError-Message, and GetExtendedErrorMessage to access the last error that occurred.

The LS:DO is available on both the Notes client and the Notes server. LS:DO is excellent for real-time data access from any Lotus-Script event in Notes, such as clicking a button, exiting a field, or opening a document. LS:DO real-time data access is the best choice for the following items:

- Optimizing data entry
  - On-the-fly lookups
  - Immediate updates
  - Input validation
  - Avoiding duplicate entries
- Mobile user queries and updates

Many developers use Notes as the data entry point for an application, which might synchronize that data with a *database management system* (DBMS) or use the DBMS for long-term data storage and archiving. The LS:DO can provide that functionality.

In addition to enabling users to issue *Structured Query Language* (SQL) statements to relational DBMSs, the LS:DO also offers data manipulation capabilities. The LS:DO supports and manages result sets, as well as provides an interface for directly using SQL when appropriate. The result set management takes the form of caching result sets, supporting navigation through the result set, and managing individual row updates—regardless of the underlying driver's cursor or ODBC conformance capabilities.

When to Use LS:DO

The LS:DO is best-suited to handle the following situations:

- LotusScript programming environment. If you develop an application with the LotusScript environment, you can easily utilize ODBC access through LS:DO classes.

- Low-volume data transfer. LS:DO is more suited for low-volume access to data resources. From a performance perspective, LS:DO is not well-suited for moving large volumes of data.

- Easy data access. If an application needs to read *and* update data in an RDBMS, LS:DO is an easier way than the ODBC API or the @DbCommands because of the classes enabling you to work with the result sets.

- Real-time direct access. LS:DO can be integrated directly into a Notes application.

LS:DO Class Library

The classes that make up the LS:DO provide you with the following benefits:

- Connection sharing. Connections are cached to avoid the added overhead of establishing a connection. In addition, because it is defined as an independent object, one connection object can be used by multiple LotusScript SQL calls.

- Multiple query and result sets. You can define multiple query objects to generate multiple result set objects, which can all be executed against the same connection and manipulated from the same script.

- Bi-directional scrolling over result set. The ODBCResultSet object provides a scrolling cursor with methods for navigating to the next, previous, first, and last rows.

- Result set search. The LocateRow method of the ODBCResultSet object provides the capacity to search for specific rows within the result set, based on specified criteria. This search capability executes faster than multiple queries or comparing values from multiple rows in LotusScript.

- Cached results. The query result in the ODBCResultSet object is optionally cached in memory (default setting) so it can be accessed later by other events in the form, increasing performance and reducing DBMS connection time. In addition, the cached result set gives you the capacity to locate records later using the Locate row method.

- Update services. Updates to back-end DBMSs through a generic ODBC interface are limited to SQL statements, where the user must ensure that the row to be updated contains a unique record reference or can be otherwise uniquely accessed through a cursor. LS:DO extends this capability by permitting individual items in a result set to be modified—without the use of an SQL statement using the SetValue method. These changes are then updated to the back-end database all at once.

- Driver transparency. Although different vendors' ODBC drivers support varying conformance levels, the LS:DO assesses these differences and often provides the same level of behavior across all drivers and databases. The developer does not have to write separate scripts for separate drivers.

ODBCConnection Class

You connect to an external data source with the ConnectTo method of ODBCConnection class. The external source must be defined in the ODBC Driver Manager at the operating system level, which stores a list of registered data source names (ODBC.INI for Windows). If the data source requires a user ID and password, pass these parameters when attempting the connection—especially with scripts running on servers or in the background where user intervention is not possible or desirable. You first can set the SilentMode property to prevent a user ID or password dialog from being started, if it is supported by your driver.

NOTE
A disconnect can take place automatically if the DisconnectTime-Out property has been set, unless transactions or queries are pending. If transactions are pending, a query result is still being retrieved, or another activity is taking place. A connection cannot be

broken with the DisconnectTimeOut property. Also, when you use a connection object in more than one query at a time, a disconnect of that
connection object will disconnect all concurrent queries.

Syntax

To create a new ODBCConnection object, use the New method.

```
Dim variableName As New ODBCConnection
```

or

```
Set odbcConnection = New ODBCConnection
```

ConnectTo Method

Connects to a data source.

Syntax

```
status = odbcConnection. ConnectTo( dataSourceName$ [ . userID$ .
password$ ] )
```

Parameters

dataSourceName$ **String.**   The ODBC name of the data source to
which you want to connect.

userID$ **String.**   Optional. The name of the user ID.

password$ **String.**   Optional. The password for the user ID.

Return Value

True = 1   Connected successfully.

False = 0   Failed to connect.

NOTE
Use this method to establish a connection to a data source. The
source is required to be already registered in the ODBC Driver
Manager.

ConnectTo Method Example

```
Sub Initialize
'Declarations
    Dim con As New ODBCConnection
'Connect to the employee data
    con.ConnectTo("EmployeeData")
End Sub
```

To connect to a data source, you must use the ConnectTo method. This example uses that method. The *EmployeeData* source is a predefined data source, defined using the ODBC Driver Manager.

ODBCQuery Class

The ODBCQuery class represents the ODBC data access features for defining a SQL statement. A query must pertain to a valid connection before it can be used or validated.

TROUBLESHOOTING TIPS

- The Query object must be linked with a data source connection. Set the Connection property of the ODBCQuery class equal to the ODBCConnection object identifier.

- The SQL property can be set to any valid SQL statement, including inserts and createtable, as well as select statements.
- The SQL statement must be in the form of a string, and may contain linked variables. For example: `"SELECT * FROM " & tablename$ & " WHERE lname='" & lastname$ & "'"`
- Strings within SQL statements require single quotations. For example: `"SELECT * FROM CUSTOMER WHERE STATE =' MA' "`
- If the SQL statement does not have parameters, it is checked for syntax errors. Otherwise, the syntax is not checked until the parameters are defined.
- You can use the NumParameters method of ODBCResultSet to get the number of parameters in the last query, if it uses parameters. You can use SetParameter to set individual parameter values and GetParameter to read them.
- The Query object extends the SQL standard query to include parameter names, for example: `"SELECT * FROM tablename WHERE date = ? AND amount = ?` can be defined as `"SELECT * FROM tablename WHERE date = ?cutoffdate? AND amount = ?checkamount?`. Use GetParameterName to retrieve these names.
- Use QueryExecuteTimeOut to prevent a query from executing for a long period of time, such as when you do not need the entire result set.

Syntax
To create a new ODBCQuery object, use the New method.

```
Dim variableName As New ODBCQuery
```

or

```
Set odbcQuery = New ODBCQuery
```

SQL Property

Read-write: Any SQL statement.

Defined in: ODBCQuery.

Data type: String.

Syntax

To get: statement$ = odbcQuery.SQL

To set: odbcQuery.SQL= statement$

> **NOTE**
> When an SQL statement is stored, it is automatically checked for
> syntactical correctness. There is no practical limit to the length of
> the SQL statement written, nor is the SQL statement limited to a
> query (select statement). The SQL statement is not executed until an
> ODBCResultSet object is created and Execute is performed. SQL state-
> ment execution depends on a user's access rights. A user may not be
> authorized to alter certain fields or tables, perform certain operations,
> or access certain tables or columns.

SQL Property Example

```
Sub Initialize
'Declarations
    Dim con As New ODBCConnection
    Dim qry As New ODBCQuery
'Connect to the employee data
    con.ConnectTo("EmployeeData")
'Set the SQL property to a valid SQL statement
    qry.SQL = "SELECT * FROM employee"
End Sub
```

ODBCResultSet Class

The ODBCResultSet class represents the ODBC data access features
for performing operations on a result set.

Syntax

To create a new ODBCResultSet object, use the New method.

```
Dim variableName As New ODBCResultSet
```

or

```
Set odbcResultSet = New ODBCResultSet
```

TROUBLESHOOTING TIP

- The ResultSet object must have a valid, properly connected Query object before a result set can be produced by the Execute method. It is strongly recommended that you check the return status for the Execute method to verify that the SQL statement was properly executed, because some SQL statements do not create result sets. For example

```
DIM Con AS NEW ODBCConnection
DIM Qry AS NEW ODBCQuery
DIM Result AS NEW ODBCResultSet
. . .
SET Qry.Connection = Con
SET Result.Query = Qry
If Result.Execute Then
( . . . use the data . . . )
Else
MessageBox("Statement did not execute")
Endif
```

- The results of the SQL statement are retrieved and cached in the ResultSet object, which enables you to use some of the result set while the rest is transmitted through the network. To accelerate the process, use FetchBatchSize to set the number of rows retrieved from the result set at a time.

- The result set is maintained in the cache as a set of rows and columns. Therefore, you can navigate through the result set using the navigation methods such as FirstRow and NextRow. Accessing rows not in memory results in a row fetch, and because result sets are usually available only sequentially, all the rows between the last cached row and the requested row must also be retrieved.

- Use GetValue to retrieve values in the columns of a result set. You may use the column's name or its ordinal value to access its value. Similarly, you can use FieldName, FieldInfo, FieldSize, and FieldNativeDataType to obtain information about a column's name at the data source, its table, its size, or its data format, respectively.

- Use SetValue to update values in rows and columns in the result set. Set the value in the column in the result set using SetValue, then update the row in the external data source using UpdateRow.

- Updating a row using UpdateRow begins a transaction on transaction-based databases. All subsequent Delete and UpdateRow transactions are part of the same transaction. You may then commit or roll-back changes using the Close or Transactions methods. Alternatively, you can set a transaction database to AutoCommit, where any change that is made is automatically committed.

- Use the special value DB_NULLVALUE to set a value to Null in the external database. For example: Call ResSet.SetValue(4, DB_NULLVALUE).

- Use HasRowChanged to detect whether other users made concurrent changes to the external database source. If the database supports ODBC cursors, you can reliably distinguish and resolve conflicts.

- Use AddRow to open a workspace for a new row in the result set object, SetValue to put the data into the fields, and UpdateRow to send the update to the external data source. Log each new row with UpdateRow before adding the next. You must have appropriate permission to update the database.

- A result set with no rows of data is valid, but many methods will not work. A result set with no rows of data is useful, because it has all the column information needed to add rows.

- Many methods and properties have associated events. *Before* events are raised after all arguments are validated locally and the action is about to be attempted. *After* events are raised after the action completes successfully, and they are not raised if the action fails.

ODBCResultSet Example

```
Sub Initialize
'Declarations
      Dim con As New ODBCConnection
      Dim qry As New ODBCQuery
'Connect to the employee data
      con.ConnectTo("EmployeeData")
'Set the SQL property to a valid SQL statement
      qry.SQL = "SELECT * FROM employee"
'Relate the result set with the ODBCQuery object
      Set result.Query = qry
'Execute the query and return a result set
      Call result.Execute()
End Sub
```

Retrieving Values from an ODBCResultSet

To retrieve values from an ODBCResultSet object, use the GetValue method.

GetValue Method

The GetValue method retrieves the value of the column referenced by *column_id%* and copies the results to the *variable* field.

Defined in: ODBCResultSet

Syntax

```
variable = odbcResultSet.GetValue( column_id%  [ , variable ] )
```

or

```
variable = odbcResultSet.GetValue(column_Name$  [ , variable ] )
```

Parameters

column_id% **Integer.** The ID of a column.

column_Name$ **String.** The name of a column.

variable Name of variable to be set. Providing the name as an argument enables the method to get its data type.

Return Value

variable Name of variable to be set.

> **NOTE**
> The data type of the returned data is determined by the data type of the second parameter. If this parameter is omitted, the data type is as set for this column by the FieldExpectedDataType property. For example, you can set the expected value to return a real value for an integer numeric field. If neither of the preceding items are in effect, the data type is closest to the native data type.

> **TROUBLESHOOTING TIP**
> Null values require special consideration when using GetValue and SetValue. To test for a null value, use the IsValueNull method. If the argument data type is a variant, a null column value is returned as an empty value. Otherwise, the data returned by GetValue is converted to a LotusScript data type. For integers and reals, the returned value is zero; for strings, the returned value is a null (zero length) string; for date-time values, the returned value is some form of *zero* time.

GetValue Example

```
Sub Initialize
'Declarations
    Dim con As New ODBCConnection
    Dim qry As New ODBCQuery
    Dim EmpName As String
'Connect to the employee data
    con.ConnectTo("EmployeeData")
'Set the SQL property to a valid SQL statement
    qry.SQL = "SELECT * FROM employee"
'Initialize the result set with the query
    Set result.Query = qry
'Execute the query and return a result set
    Call result.Execute()
'Get a value from the result set
    status = result.GetValue( "EmployeeName", EmpName )
End Sub
```

Inserting Values into an ODBCResultSet

To insert values into an ODBCResultSet, use the SetValue method.

SetValue Method

The ODBCResultSet class SetValue method sets a new value to a column.

Syntax

```
status = odbcResultSet.SetValue( column_id% , value )
```

or

```
status = odbcResultSet.SetValue( column_Name$ , value )
```

Parameters

column_id% **Integer.**   The column ID.

column_Name$ **String.**   The column name.

value   A value of a type consistent with the column.

Return Value

True   Current value in specified column is updated by *value*.

False   Value cannot be changed, similar to when the result set is read-only.

> **NOTE**
> The data is automatically converted to the required native data type. The changes take effect when an UpdateRow operation occurs; however, in transaction mode, a commit operation must also occur. Specify the *value* argument as DB_NULLVALUE to set a null value of the correct type.

SetValue Example

```
Sub Initialize
'Declarations
      Dim con As New ODBCConnection
      Dim qry As New ODBCQuery
      Dim EmpName As String
      Dim EmpSalary As Double
'Connect to the employee data
      con.ConnectTo("EmployeeData")
'Set the SQL property to a valid SQL statement
      qry.SQL = "SELECT * FROM employee"
'Initialize the result set with the query
      Set result.Query = qry
```

```
'Execute the query and return a result set
     Call result.Execute()
'Get a value from the result set
     EmpName = "Roland Lim"
     EmpSalary = 1000000
     status = Result.SetValue( "EmployeeName", EmpName )
     status = Result.SetValue( "Salary", EmpSalary )
End Sub
```

Sample Questions

1. *Objective:* LS:DO

   Which class is not included in the LS:DO?

   a. ODBCConnection

   b. ODBCResultSet

   c. ODBCGetValue

   d. ODBCQuery

2. *Objective:* LS:DO

   What line must you include in the script Options when using the LS:DO?

   a. `Use "LSDO"`

   b. `UseLSX "*LSXODBC"`

   c. `Use_SX "*LSDO"`

   d. `UseLSX "LSXODBC"`

3. *Objective:* LS:DO

   The LS:DO is best-suited to handle

   a. Data transfers

   b. High-volume data transfers

   c. Document transfers

   d. Low-volume data transfers

4. *Objective:* LS:DO class library

   The LotusScript:Data Object provides you with which benefit?

   a. Connection sharing

   b. Multiple query and result sets

   c. Result set search

   d. All of the above

5. *Objective:* ODBCConnection class

   To connect to an external data source, you

   a. Should use the ConnectToODBC method

   b. Need to have the external source defined in the ODBC Driver Manager

   c. Do not need to have the external source defined in the ODBC Driver Manager

   d. Should use the Connect method

6. *Objective:* ODBCQuery class

   Juan is trying to connect to an external database using the following script. Which line is causing him trouble?

   ```
   Dim con As New ODBCConnection
   Dim qry As New ODBCQuery
   Dim result As New ODBCResultSet
   con.ConnectTo("EmployeeData")-Line 1
   SQL = "SELECT * FROM employee"-Line 2
   Set result.Query = qry-Line 3
   Call result.Execute()-Line 4
   ```

   a. Line 1

   b. Line 2

   c. Line 3

   d. Line 4

7. *Objective:* ODBCResultSet class

   True or False: The ODBCResultSet object must have a valid, properly connected ODBCQuery object before a result set can be produced by the Run method.

8. *Objective:* Retrieving values from a result set

   The second parameter in the GetValue method

   a. Gets the value from the result set

   b. Is the variable in LotusScript to be set

   c. Is the name of the column from which to retrieve the value

   d. There is no second parameter.

9. *Objective:* Assigning values to a result set

   When assigning a value to a result set

   a. The value is set and is immediately passed to the non-Notes source database.

   b. The value is not passed onto the non-Notes source database until a Commit is executed.

   c. The value is set directly in the non-Notes source database.

   d. The value is not passed onto the non-Notes source database until an UpdateRow is executed.

Sample Answers

1. *Answer:* c

   The LS:DO consists of a set of three classes: ODBCConnection, ODBCQuery, and ODBCResultSet.

2. *Answer:* b

   You must put the following statement in the (Options) event of the (Global) object to access the OBDC classes: `UseLSX "*LSXODBC"`. The leading * tells LotusScript to use the class registry to look up the path of the LS:DO dynamic library being loaded. This method is a platform-independent way of loading LS:DO, because each operating system uses different methods.

3. *Answer:* d

   The LS:DO is best-suited to handle the following situations:

   - LotusScript programming environment. If you develop an application with the LotusScript environment, you can easily utilize ODBC access through LS:DO classes.
   - Low-volume data transfer. LS:DO is more suited for low-volume access to data resources. From a performance perspective, LS:DO is not well-suited for moving large volumes of data.
   - Easy data access. If an application needs to read and update data in an RDBMS, LS:DO is an easier way than the ODBC API or the @DbCommands because of the classes enabling you to work with the result sets.
   - Real-time direct access. LS:DO can be integrated directly into a Notes application.

4. *Answer:* b

5. *Answer:* b

   You connect to an external data source with the ConnectTo method of ODBCConnection class. The external source must be defined in the ODBC Driver Manager at the operating system level, which stores a list of registered data source names (`ODBC.INI` for Windows).

6. *Answer:* c

   You must set the ODBCQuery object SQL property for the query to execute properly.

7. *Answer:* False

   The ResultSet object must have a valid, properly connected Query object before a result set can be produced by the Execute method.

8. *Answer:* b

   The second parameter is the name of variable to be set. Providing the name as an argument enables the method to get its data type.

9. *Answer:* b

   The changes take effect when an UpdateRow operation occurs; however, in transaction mode, a Commit operation must also occur.

CHAPTER 11

Notes Classes

NotesACL Class

Represents the ACL of a database.

Containment

- Contained by NotesDatabase.
- Contains NotesACLEntry.
- See Table 11.1 for more information.

Table 11.1 NotesACL

Properties	
Parent	Read-only. The database that owns an ACL
Roles	Read-only. All the roles defined in an ACL
UniformAccess	Read-write. Indicates whether the uniform access is set
Methods	
AddRole	Adds a role to an ACL
CreateACLEntry	Creates an entry in the ACL with the name and level that you specify. When used with OLE automation, this method enables you to create a NotesACLEntry object without using New
DeleteRole	Deletes a role from an ACL
GetEntry	Given a name, finds its entry in an ACL
GetFirstEntry	Returns the first entry in an ACL, usually the Default entry
GetNextEntry	Given an ACL entry, returns the next one
RenameRole	Changes the name of a role
Save	Saves the changes you have made to the ACL. If you do not call Save before closing a database, the changes you have made to its ACL are lost

NotesACLEntry Class

Represents a single entry in an ACL. An entry may be for a person, a group, or for a server.

Containment

- Contained by NotesACL. See Table 11.2 for more information.

Table 11.2   NotesACLEntry

Properties

CanCreateDocuments	Read-Write. For an entry with Author access to a database, indicates whether the entry is enabled to create new documents
CanCreatePersonalAgent	Read-write. Indicates whether an entry can create personal agents in a database
CanCreatePersonalFolder	Read-write. Indicates whether an entry can create personal folders in a database
CanDeleteDocuments	Read-Write. For an entry with Author access or higher to a database, indicates whether an entry can delete documents
IsPublicReader	Read-write. Indicates whether the current entry is a public reader of the database
IsPublicWriter	Read-write. Indicates whether the current entry is a public writer of the database
Level	Read-Write. The access level this entry has for this database
Name	Read-Write. The name of an entry
Parent	Read-only. The ACL that contains an entry
Roles	Read-only. The roles that are enabled for an entry

Methods

DisableRole	Given a role, disables the role for an entry

continues

Table 11.2 Continued

Methods	
EnableRole	Given the name of a role, enables the role for an entry
IsRoleEnabled	Given a role, indicates whether the role is enabled for an entry
New	Creates an entry in an ACL with the name and level that you specify
Remove	Removes an entry from an ACL

NotesAgent Class

Represents an agent. The agent might be public or personal and might be a R3 macro.

Containment

- Contained by NotesSession and NotesDatabase. See Table 11.3 for more information.

Table 11.3 NotesAgent

Properties	
Comment	Read-write. Any comment
CommonOwner	Read-only. The name of the person who last modified and saved an agent (the agent's owner). If the owner has a hierarchical name, only the common name is returned.
IsEnabled	Read-only. Indicates whether an agent is able to run

Properties	
IsPublic	Read-only. Indicates whether an agent is public or personal
	■ A public agent is accessible to all users of a database and is stored in the database.
	■ A personal agent is accessible only to its owner and is stored in the owner's desktop file.
LastRun	Read-only. The date that an agent last ran
Name	Read-only. The name of an agent. Within a database, the name of an agent may not be unique.
Owner	Read-only. The name of the person who last modified and saved an agent (the agent's owner)
Parent	Read-only. The database that contains an agent
Query	Read-only. The text of the query used by an agent to select documents. In the Agent Builder, a query is defined by the searches added to the agent using the Add Search button.
ServerName	Read-only. The name of the server on which an agent runs
Method	
Remove	Permanently deletes an agent from a database
Run	Runs the agent

NotesDatabase Class

Represents a Notes database.

Containment

- Contained by NotesSession, NotesDbDirectory, and NotesUI-Database.

- Can contain NotesACL, NotesAgent, NotesDocument, Notes-DocumentCollection, NotesForm, and NotesView. See Table 11.4 for more information.

Table 11.4 NotesDatabase

Properties	
ACL	Read-only. The ACL for a database
Agents	Read-only. All of the agents in a database
AllDocuments	Read-only. All the documents in a database
Categories	Read-Write. The categories under which a database appears in the Database Library. Multiple categories are separated by a comma or semicolon.
Created	Read-only. The date a database was created
CurrentAccessLevel	Read-only. The current user's access level to a database
DelayUpdates	Read-write. Indicates whether updates to a server are delayed (batched) for better performance
DesignTemplateName	Read-only. The name of the design template from which a database inherits its design. If the database does not inherit its design from a design template, returns an empty string (" ").
FileName	Read-only. The filename of a database, excluding the path and .nsf extension
FilePath	Read-only. The path and filename of a database. The filename includes the .nsf extension.
Forms	Read-only. All the forms in a database

Properties	
IsFTIndexed	Read-only. Indicates whether a database has a full-text index
IsMultiDbSearch	Read-only. Indicates whether a database represents a multi-database search index
IsOpen	Read-only. Indicates whether a database is open
IsPrivateAddressBook	Read-only. Indicates whether a database is a Personal Address Book
IsPublicAddressBook	Read-only. Indicates whether a database is a Public Address Book
LastFTIndexed	Read-only. The date when a database's full-text index was last updated
LastModified	Read-only. The date that a database was last modified
Managers	Read-only. People, servers, and groups that have Manager access to a database
Parent	Read-only. The Notes session that contains a database
PercentUsed	Read-only. The percentage of a database's total size that is occupied by real data (versus empty space)
ReplicaID	Read-only. A 16-character combination of letters and numbers that represents the replica ID of a Notes database. Any databases with the same replica ID are replicas of one another.
Server	Read-only. The name of the server where a database resides
Size	Read-only. The size of a database, in bytes
SizeQuota	Read-write. The size quota of a database, in KB

continues

Table 11.4 Continued

Properties	
TemplateName	Read-only. The template name of a database, if the database is a template. If the database is not a template, returns an empty string (" ").
Title	Read-Write. The title of a database
UnprocessedDocuments	Read-only. The documents in a database that the current agent or view action considers *unprocessed*. The type of agent determines which documents are considered unprocessed.
Views	Read-only. The views and folders in a database
Methods	
Compact	Compacts a local database
Create	Creates a new database on disk, using the server and filename that you specify. Because the new database is not based on a template, it is blank and does not contain any forms or views.
CreateCopy	Creates an empty copy of the current database. The copy contains the design elements of the current database, an identical ACL, and an identical title. The copy does not contain any documents and is not a replica.
CreateDocument	Creates a document in a database and returns a NotesDocument object that represents the new document. You must call Save if you want the new document to be saved to disk. When used with OLE automation, this method enables you to create a NotesDocument object without using New.

Methods

CreateFromTemplate	If the current database is a template, creates a new database from the template. The new database has the design features and documents of the template.
CreateReplica	Creates a replica of the current database at a new location. The new replica has an identical ACL.
FTSearch	Conducts a full-text search of all the documents in a database
GetAgent	Finds an agent in a database, given the agent name
GetDocumentByID	Finds a document in a database, given the document NoteID
GetDocumentByUNID	Finds a document in a database, given the document universal ID (UNID)
GetDocumentByURL	Instantiates a document in the Web Navigator database and returns a NotesDocument object for it. You can use this method for either the Server Web Navigator or Personal Web Navigator databases.
GetForm	Finds a form in a database, given the form name
GetProfileDocument	Retrieves or creates a profile document
GetURLHeaderInfo	Gets the specific *HyperText Transport Protocol* (HTTP) header information from the *Uniform Resource Locator* (URL). A URL is a text string used for identifying and addressing a Web resource.
GetView	Finds a view or folder in a database, given the name or alias of the view or folder

continues

Table 11.4 Continued

Methods	
GrantAccess	Modifies a database ACL to provide the specified level of access to a person, group, or server
New	Creates a NotesDatabase object that represents the database located at the server and filename you specify and opens the database, if possible. Unlike the behavior of New in other classes (such as the NotesDocument), using New for a NotesDatabase does *not* create a new database on disk.
Open	Opens a database. A database must be open for a script to access its properties and methods.
OpenByReplicaID	Given a server name and a replica ID, opens the specified database, if it exists
OpenIfModified	Given a date, opens the specified database if it has been modified since that date
OpenMail	Assigns a database to the current user's mail database and opens the database
OpenURLDb	Finds and opens the default Web Navigator database
OpenWithFailover	Opens a database on a server
QueryAccess	Returns a person's, group's, or server's current access level to a database
Remove	Permanently deletes a database from disk
Replicate	Replicates a database with its replica(s) on a given server

Methods	
RevokeAccess	Removes a person, group, or server from a database ACL. This command resets the access level for that person, group, or server to the Default setting for the database.
Search	Given selection criteria for a document, returns all documents in a database that meet the criteria
UnprocessedFTSearch	Given selection criteria for a document, returns documents in a database that ■ The current agent considers to be unprocessed ■ Match the query
UnprocessedSearch	Given selection criteria for a document, returns documents in a database that ■ The current agent considers to be unprocessed ■ Meet the criteria ■ Were created or modified since the cutoff date
UpdateFTIndex	Updates the database's full-text index

NotesDateRange Class

Represents a range of dates and times.

Containment

■ Contained by NotesSession.

■ Contains NotesDateTime. See Table 11.5 for more information.

Table 11.5 NotesDataRange

Properties	
StartDateTime	Read-write. The starting date-time of a range
EndDateTime	Read-write. The ending date-time of a range
Text	Read-write. The text associated with a range

NotesDateTime Class

Represents a date and time. Provides a means of translating between the LotusScript date-time format and the Notes format.

Containment

- Contained by NotesDateRange, NotesSession. See Table 11.6 for more information.

Table 11.6 NotesDateTime

Properties	
GMTTime	Read-only. A string representing a date-time, converted to Greenwich Mean Time (time zone zero)
IsDST	Read-only. Indicates whether the time reflects Daylight Savings Time
LocalTime	Read-Write. A string representing a date-time, in the local time zone
LSGMTTime	Read-only. A LotusScript variant representing a date-time, converted to Greenwich Mean Time (time zone zero)
LSLocalTime	Read-Write. A LotusScript variant representing a date-time, in the local time zone

Properties

TimeZone	Read-only. An integer representing the time zone of a date-time. In many cases, but not all, this integer indicates the number of hours which must be added to the time to get Greenwich Mean Time. May be positive or negative
ZoneTime	Read-only. A string representation of the time adjusted for the TimeZone and IsDST properties

Methods

AdjustDay	Increments a date-time by the number of days you specify
AdjustHour	Increments a date-time by the number of hours you specify
AdjustMinute	Increments a date-time by the number of minutes you specify
AdjustMonth	Increments a date-time by the number of months you specify
AdjustSecond	Increments a date-time by the number of seconds you specify
AdjustYear	Increments a date-time by the number of years you specify
ConvertToZone	Changes the TimeZone and IsDST properties as specified
New	Given a string that represents the date and time you want, New creates an object that represents that date and time.
SetAnyDate	Sets the date component to a wild card value, which means it will match any date. The time component is unaffected.
SetAnyTime	Sets the time component to a wild card value, which means it will match any time. The date component is unaffected.

continues

Table 11.6 Continued

Methods	
SetNow	Sets the value of a date-time to now (today's date and current time)
TimeDifference	Finds the difference in seconds between one date-time and another

NotesDbDirectory Class

Represents the Notes databases on a specific server or local computer.

Containment

- Contained by NotesSession. Contains: NotesDatabase. See Table 11.7 for more information.

Table 11.7 NotesDbDirectory

Property	
Name	Read-only. The name of the server whose database directory you are searching. This property is set when you create a database directory using New.

Methods	
GetFirstDatabase	Returns the first database on a server (or local computer), using the file type you specify
GetNextDatabase	Returns the next database in a directory, using the file type specified in the GetFirstDatabase method
New	Creates a new NotesDbDirectory object using the name of the server you want to access

NotesDocument Class

Represents a document in a database.

Containment

- Contained by NotesDatabase, NotesDocumentCollection, Notes-Newsletter, NotesUIDocument, and NotesView.
- Contains NotesEmbeddedObject, NotesItem, NotesRichTextItem. See Table 11.8 for more information.

Table 11.8   NotesDocument

Properties	
Authors	Read-only. The names of the people who have saved a document
ColumnValues	Read-only. An array of values, each element of which corresponds with a column value in the document's Parent view. The first value in the array is the value that appears in the view's first column for the document, the second value is the one that appears in the second column, and so on. The value of each element of the array is the result of the corresponding column's formula and the items on the current document. Some elements in the array might have no value.
Created	Read-only. The date a document was created
EmbeddedObjects[1]	Read-only. The OLE/2 and OLE/1 embedded objects in a document
EncryptionKeys	Read-Write. The key(s) used to encrypt a document. The Encrypt method uses these keys when it encrypts the document.

continues

Table 11.8 Continued

Properties	
EncryptOnSend	Read-Write. Indicates whether a document is encrypted when mailed
FTSearchScore	Read-only. The full-text search score of a document, if it was retrieved as part of a full-text search
HasEmbedded[2]	Read-only. Indicates whether a document contains one or more embedded objects, object links, or file attachments
IsNewNote	Read-only. Indicates whether a document is new. A document is new if it has not been saved.
IsProfile	Read-only. Indicates whether a Notes-Document object is a profile document
IsResponse	Read-only. Indicates whether a document is a response to another document
IsSigned	Read-only. Indicates whether a document contains a signature
IsUIDocOpen	Read-only. Indicates whether a NotesDocument object (back-end document) is accessed through a NotesUIDocument object (front-end document)
Items	Read-only. All the items on a document. An item is any piece of data stored in a document.
Key	Read-only. If a profile document, the username (key) of the profile
LastAccessed	Read-only. The date a document was last modified or read
LastModified	Read-only. The date a document was last modified

Properties

NameOfProfile	Read-only. If a profile document, the name of the profile
NoteID	Read-only. The NoteID of a document, which is an eight-character combination of letters and numbers that uniquely identifies a document within a particular database
ParentDatabase	Read-only. The database that contains a document
ParentDocumentUNID	Read-only. The universal ID of a document's parent, if the document is a response. Returns an empty string (" ") if a document does not have a parent
ParentView	Read-only. The view from which a document was retrieved, if any. If the document was retrieved directly from the database or a document collection, returns Nothing
Responses	Read-only. The immediate responses to a document
SaveMessageOnSend	Read-Write. Indicates whether a document is saved to a database when mailed. Only applies to new documents that have not yet been saved
SentByAgent	Read-only. Indicates whether a document was mailed by a script
Signer	Read-only. The name of the person who created the signature, if a document is signed
SignOnSend	Read-Write. Indicates whether a document is signed when mailed
Size	Read-only. The size of a document in bytes, which includes the size of any file attachments on the document

continues

Table 11.8 Continued

Properties	
UniversalID	Read-write. The universal ID of a document, which is a 32-character combination of hexadecimal digits (0–9, A–F) that uniquely identifies a document across all replicas of a databases. If two documents in replica databases share the same universal ID, the documents are replicas.
Verifier	Read-only. The name of the certificate that verified a signature, if a document is signed
Methods	
AppendItemValue	Creates a new item on a document and sets the item value
ComputeWithForm	Validates a document by executing the default value, translation, and validation formulas, if any are defined in the document form
CopyAllItems	Given a destination document, copies all of the items in the current document into the destination document. The item names are unchanged
CopyItem	Given an item, copies it into the current document and optionally assigns the copied item a new name
CopyToDatabase	Copies a document into the specified database
CreateReplyMessage	Creates a new document that is formatted as a reply to the current document

Methods

CreateRichTextItem	Creates a new rich text item on a document using a name you specify and returns the corresponding NotesRichTextItem object. When used with OLE automation, this method enables you to create a new rich text item and NotesRichTextItem object without using New.
Encrypt	Encrypts a document in a database
GetAttachment	Given the name of a file attachment, returns a NotesEmbeddedObject representing the attachment. You can use this method to find file attachments which are not contained in a rich text item (such as an attachment in a R2 database), as well as file attachments that are contained in a rich text item.
GetFirstItem	Given a name, returns the first item of the specified name belonging to the document
GetItemValue	Given the name of an item, returns the value of that item on a document
HasItem	Given the name of an item, indicates whether that item exists on the document
MakeResponse	Makes one document a response to another document. The two documents must be in the same database.
New	Given a database, New creates a document in the database and returns a NotesDocument object that represents the document. You must call Save if you want the new document to be saved to disk.
PutInFolder	Adds a document to the specified folder. If the folder does not exist in the document's database, it is created.

continues

Table 11.8 Continued

Methods	
Remove	Permanently deletes a document from a database
RemoveFromFolder	Removes a document from the specified folder
RemoveItem	Given the name of an item, deletes the item from a document
RenderToRTItem	Creates a picture of a document and places it into a rich text item you specify. The picture is created using both the document and its form; therefore, the form's input translation and validation formulas are executed.
ReplaceItemValue	Replaces all items of the specified name with one new item, which is assigned the specified value. If the document does not contain an item with the specified name, the method creates a new item and adds it to the document.
Save	Saves any changes you have made to a document
Send	Mails a document to the recipients you specify
Sign	Signs a document

[1]EmbeddedObjects is not supported under OS/2, UNIX, and on the Macintosh.
[2]HasEmbedded is not supported under OS/2, UNIX, and on the Macintosh.

NotesDocumentCollection Class

Represents a collection of documents from a database, selected according to specific criteria.

Containment

- Contained by NotesDatabase, NotesSession, NotesUIDatabase, NotesUIView
- Contains NotesDocument. See Table 11.9 for more information.

Table 11.9   NotesDocumentCollection

Properties	
Count	Read-only. The number of documents in a collection
IsSorted	Read-only. Indicates whether the documents in a collection are sorted. A collection is sorted only when it results from a full-text search of a database.
Parent	Read-only. The database that contains a document collection
Query	Read-only. The text of the query that produced a document collection, if the collection results from a full-text or other search
Methods	
FTSearch	Conducts a full-text search of all the documents in a Notes database collection, and reduces the collection to those documents that match
GetFirstDocument	Gets the first document in a collection
GetLastDocument	Gets the last document in a collection
GetNextDocument	Given a document, finds the document immediately following it in a collection
GetNthDocument	Given a position number, returns the document at that position in a collection
GetPrevDocument	Given a document, finds the document immediately preceding it in a collection

continues

Table 11.9   Continued

Methods	
PutAllInFolder	Adds all the documents in the collection to the specified folder. If the folder does not exist in the document's database, it is created.
RemoveAll	Permanently deletes all documents in a collection from a database
RemoveAllFromFolder	Removes all documents in the collection from the specified folder
StampAll	Replaces the value of a specified name in all documents in a collection
UpdateAll	Marks all documents in a collection as processed by an agent

NotesEmbeddedObject Class

Represents any one of the following

- An embedded object
- An object link
- A file attachment

Some methods and properties that are available for embedded and linked objects are unavailable for file attachments.

 NOTE
NotesEmbeddedObject is not supported under OS/2, UNIX, and on the Macintosh.

Containment

- Contained by NotesDocument and NotesRichTextItem
- Table 11.10 shows more information.

Table 11.10 NotesEmbeddedObject

Properties	
Class	Read-only. The name of the application which created an object
FileSize	Read-only. The size of an embedded object, object link, or file attachment, in bytes
Name	Read-only. The name used to reference an embedded object or object link
Object	Read-only. If an embedded object has been loaded into memory, returns the OLE handle (IUnknown or IDispatch handle). If the OLE object supports OLE Automation, you can invoke the methods and properties of the object using the handle.
Parent	Read-only. The rich text item that holds an object
Source	Read-only ■ If the NotesEmbeddedObject is an embedded object or object link, this property returns the internal name that Notes uses to refer to the source document. ■ If the NotesEmbeddedObject is a file attachment, this property returns the filename of the original file.
Type	Read-only. Indicates whether a NotesEmbeddedObject is an embedded object, an object link, or a file attachment
Verbs	Read-only. The verbs that an object supports, if the object is an OLE/2 embedded object
Methods	
Activate	Causes an embedded object or object link to be loaded by OLE
DoVerb	Given the name of a verb, executes the verb in an embedded object

continues

Table 11.10 Continued

Methods	
ExtractFile	Copies a file attachment to disk
Remove	Permanently deletes an embedded object, object link, or file attachment

NotesForm Class

Represents a form in a database.

Containment

- Contained by NotesDatabase
- Table 11.11 shows more information.

Table 11.11 NotesForm

Properties	
Aliases	Read-only. The aliases of a form
Fields	Read-only. The names of all the fields of a form
FormUsers	Read-write. The contents of the $FormUsers field
IsSubForm	Read-only. Indicates whether form is a subform
Name	Read-only. The name of a form
ProtectReaders	Read-write. Protects $Readers items from being overwritten by replication
ProtectUsers	Read-write. Protects $FormUsers items from being overwritten by replication
Readers	Read-write. The contents of the $Readers field
Methods	
Remove	Permanently deletes a form from a database

NotesInternational Class

Represents the international settings in the operating environment, for example, the Windows Control Panel international settings. When any of these settings is changed in the operating environment, Notes immediately recognizes the new settings.

Containment

- Contained by NotesSession. Table 11.12 shows more information.

Table 11.12   NotesInternational

Properties	
AMString	Read-only. The string that denotes AM time, for example, *AM* in English
CurrencyDigits	Read only. Indicates the number of decimal digits in number format
CurrencySymbol	Read-only. The symbol that indicates a number is currency, for example, the dollar sign ($)
DateSep	Read-only. The character used to separate months, days, and years, for example, the slash (/)
DecimalSep	Read-only. The decimal separator in number format, for example, the decimal point (.)
IsCurrencySpace	Read-only. Indicates whether the currency format has a space between the currency symbol and the number
IsCurrencySuffix	Read-only. Indicates whether the currency symbol follows the number in the currency format
IsCurrencyZero	Read-only. Indicates whether fractions have a zero before the decimal point in number format

continues

Table 11.12   Continued

Properties	
IsDateDMY	Read-only. Indicates whether the order of the date format is day-month-year
IsDateMDY	Read-only. Indicates whether the order of the date format is month-day-year
IsDateYMD	Read-only. Indicates whether the order of the date format is year-month-day
IsDST	Read-only. Indicates whether the time format reflects Daylight Savings Time
IsTime24Hour	Read-only. Indicates whether the time format is 24-hour
PMString	Read-only. The string that denotes PM time, for example, *PM* in English
ThousandsSep	Read only. The thousands separator in number format, for example, the comma (,)
TimeSep	Read-only. The character used to separate hours, minutes, and seconds, for example, the colon (:)
TimeZone	Read-only. An integer representing the time zone. In many cases, but not all, this integer indicates the number of hours which must be added to the time to get Greenwich Mean Time. Can be positive or negative
Today	Read-only. The string that means today in a time-date specification, for example, *Today* in English
Tomorrow	Read-only. The string that means tomorrow in a time-date specification, for example, *Tomorrow* in English
Yesterday	Read-only. The string that means yesterday in a time-date specification, for example, *Yesterday* in English

NotesItem Class

This class represents a piece of data in a document.

In the user interface, Notes displays items in a document through fields on a form. When a field on a form and an item in a document have the same name, the field displays the item (for example, the Subject field displays the Subject item).

All of the items in a document are accessible through LotusScript, regardless of which form is used to display the document in the user interface.

Derived Classes

- NotesRichTextItem inherits from the NotesItem class.

Containment

- Contained by NotesDocument. Table 11.13 shows more information.

Table 11.13   NotesItem

Properties	
DateTimeValue	Read-Write. For a date-time item, returns a NotesDateTime object representing the value of the item. For items of other types, returns Nothing
IsAuthors	Read-write. Indicates whether an item is of type Authors. An Authors item contains a list of Notes usernames, indicating those who have Author access to a particular document.
IsEncrypted	Read-Write. Indicates whether an item is encrypted
IsNames	Read-write. Indicates whether an item is a Names item. A Names item contains a list of Notes usernames.

continues

Table 11.13 Continued

Properties	
IsProtected	Read-Write. Indicates whether a user needs at least Editor access to modify an item
IsReaders	Read-Write. Indicates whether or not an item is of type Readers. A Readers item contains a list of Notes usernames, indicating those who have Reader access to a particular document.
IsSigned	Read-Write. Indicates whether an item contains a signature
IsSummary	Read-Write. Indicates whether an item can appear in a view or folder
LastModified	Read-only. The date that an item was last modified
Name	Read-only. The name of an item
Parent	Read-only. The document that contains an item
SaveToDisk	Read-write. Indicates whether an item is saved to disk when the document is saved
Text	Read-only. A plain text representation of an item's value
Type	Read-only. The data type of an item
ValueLength	Read-only. The size of an item's value in bytes
Values	Read-Write. The value(s) that an item holds
Methods	
Abstract	Abbreviates the contents of a text item

Methods	
AppendToTextList	For an item that is a text list, adds a new value to the item without erasing any existing values
Contains	Given a value, checks whether the value matches at least one of the item's values exactly
CopyItemToDocument	Copies an item to a specified document
New	Given a document, New creates an item on the document with a name and value that you specify. The data type of the item depends on the value you give it.
Remove	Permanently deletes an item from a document

NotesLog Class

Enables you to record actions and errors that take place during a script's execution. You can record actions and errors in

- A Notes database
- A mail memo
- A file (for scripts that run locally)
- An agent log

Containment

- Contained by NotesSession.
- Table 11.14 shows more information.

Table 11.14   NotesLog

Properties	
LogActions	Read-Write. Indicates whether action logging is enabled
LogErrors	Read-Write. Indicates whether error logging is enabled
NumActions	Read-only. The number of actions logged so far
NumErrors	Read-only. The number of errors logged so far
OverwriteFile	Read-Write. For a log that records to a file, indicates whether the log should write over the existing file or append to it. This property has no effect on logs that record to a mail message or database.
ProgramName	Read-Write. The name that identifies the script whose actions and errors you are logging. The name is the same as the name specified with New or CreateLog.
Methods	
Close	Closes a log
LogAction	Records an action in a log
LogError	Records an error in a log
LogEvent	Sends a Notes event out to the network. Only scripts running on a server can use this method.
New	Creates a new log
OpenAgentLog	Opens the agent log for the current agent
OpenFileLog	Starts logging to a specified disk file. This method returns an error if you call it on a server.
OpenMailLog	Opens a new mail memo for logging. The memo is mailed when the log's Close method is called —or when the object is deleted.
OpenNotesLog	Opens a specified Notes database for logging

NotesName Class

Represents a user or server name.

Containment

- Contained by NotesSession. Table 11.15 shows more information.

Table 11.15   NotesName

Properties	
Abbreviated	Read-only. A hierarchical name in abbreviated form
ADMD	Read-only. The *Administration Management Domain Name* (ADMD) associated with the name
Canonical	Read-only. A hierarchical name in canonical form
Common	Read-only. The common name component of a hierarchical name (CN=), or the entire name if it is flat
Country	Read-only. The country component of a hierarchical name (C=)
Generation	Read-only. The generation part of a name, for example, *Jr.*
Given	Read-only. The given part of a name
Initials	Read-only. The initials part of a name
IsHierarchical	Read-only. Indicates whether a name is hierarchical
Keyword	Read-only. The following components of a hierarchical name in the order shown separated by backslashes: country\organization\organizational unit 1\.organizational unit 2\.organizational unit 3\.organizational unit 4
Organization	Read-only. The organization component of a hierarchical name (O=)

continues

Table 11.15   Continued

Properties	
OrgUnit1	Read-only. The first organizational unit of a hierarchical name (OU=)
OrgUnit2	Read-only. The second organizational unit component of a hierarchical name (OU=)
OrgUnit3	Read-only. The third organizational unit component of a hierarchical name (OU=)
OrgUnit4	Read-only. The fourth organizational unit component of a hierarchical name (OU=)
PRMD	Read-only. The *Private Management Domain Name* (PRMD) of the name
Surname	Read-only. The surname part of the name
Methods	
New	Creates a new NotesName object

NotesNewsletter Class

A document or set of documents that contain information from, or links to, several other documents.

Containment

- Contained by NotesSession. Contains: NotesDocument Table 11.16 shows more information.

Table 11.16 NotesNewsLetter

Properties	
DoScore	Read-Write. For a newsletter document created using the FormatMsgWithDoclinks method, indicates whether the newsletter includes each document's relevance score
DoSubject	Read-Write. For a newsletter document created using the FormatMsgWithDoclinks method, indicates whether the newsletter includes a string describing the subject of each document
SubjectItemName	Read-Write. For a newsletter document created using the FormatMsgWithDoclinks method, indicates the name of the item on a newsletter's documents which contains the text you want to use as a subject line
Methods	
FormatDocument	Creates a new document in the given database containing a rendering (picture) of a specified document in the newsletter's collection. This action is similar to forwarding a document, which displays a picture of the forwarded document.
FormatMsgWithDoclinks	Creates a newsletter document in the given database that contains a link to each document in the newsletter's collection
New	Creates a new NotesNewsletter object

NotesRichTextItem Class

Represents an item of type rich text.

Base Class

- Inherits from: NotesItem.

Containment

- Contained by NotesDocument.
- Contains NotesEmbeddedObject. Table 11.17 shows more information.

Table 11.17   NotesRichTextItem

Property	
EmbeddedObjects[1]	Read-only. All the embedded objects, object links, and file attachments contained in a rich text item
Methods	
AddNewLine	Appends one or more new lines (carriage returns) to the end of a rich text item
AddTab	Appends one or more tabs to the end of a rich text item
AppendDocLink	Given a database, view, or document to link to, adds a link to the end of a rich text item
AppendRTFile	Appends the contents of a rich text file (composed of compound document records) to the end of a rich text item
AppendRTItem	Appends the contents of one rich text item to the end of another rich text item

Methods

AppendText	Appends text to the end of a rich text item. The text is rendered with the current style of the item (such as bold or italicized).
EmbedObject[1]	Given the name of a file or an application, does one of the following
	■ Attaches the file you specify to a rich text item
	■ Embeds an object in a rich text item. The object is created using either the application or the file you specify.
	■ Places an object link in a rich text item. The link is created using the file you specify.
GetEmbeddedObject[1]	Given the name of a file attachment, embedded object, or object link in a rich text item, returns the corresponding NotesEmbeddedObject
GetFormattedText	Returns the contents of a rich text item as plain text
New	Given a document, New creates a rich text item on the document with the name you specify

[1]EmbeddedObjects is not supported under OS/2, UNIX, and on the Macintosh.

NotesSession Class

Represents the Notes environment of the current script, providing access to environment variables, Address Books, information about the current user, and information about the current Notes platform and release number.

Containment

- Contains NotesAgent, NotesDatabase, NotesDateRange, Notes-DateTime, NotesDbDirectory, NotesDocumentCollection, Notes-International, NotesLog, NotesNewsletter, and NotesTimer. Table 11.18 shows more information.

Table 11.18   NotesSession

Properties	
AddressBooks	Read-only. The Address Books, both public and personal, that are known to the current script
CommonUserName	Read-only. The common name portion of the current user's name
CurrentAgent	Read-only. The agent that is currently running
CurrentDatabase	Read-only. The database in which the current script resides. This database might or might not be open.
DocumentContext	Read-only. An in-memory document created by an external program through the Notes API
EffectiveUserName	Read-only. The username that is in effect for the current script ■ For a script running on a workstation, this is the name of the current user ■ For a script running on a server, this is the name of the script's owner (the person who last saved the script)
International	Read-only. The international (regional) settings for your operating environment
IsOnServer	Read-only. Indicates whether a script is running on a server

Properties

LastExitStatus	Read-only. The exit status code returned by the Agent Manager the last time the current agent ran
LastRun	Read-only. The date when the current agent was last executed
NotesVersion	Read-only. The release of Notes in which the current script is running
Platform	Read-only. The name of the platform in which the current script is running
SavedData	Read-only. A document that an agent script uses to store information between invocations. The script can use the information in this document the next time the script runs.
UserName	Read-only. The current user's name
	■ For a script running on a workstation, this is the name of the current user
	■ For a script running on a server, this is the name of the server

Methods

CreateDateRange	Creates a new NotesDateRange object
CreateDateTime	Given a string that represents the date and time you want, creates a new NotesDateTime object representing that date and time. When used with OLE automation, this method enables you to create a NotesDateTime object without using New
CreateLog	Creates a new NotesLog object with the name you specify. When used with OLE automation, this method enables you to create a NotesLog object without using New

continues

Table 11.18 Continued

Methods	
CreateName	Creates a new NotesName object. When used with OLE automation, this method enables you to create a NotesName object without using the New method of NotesName.
CreateNewsletter	Given a NotesDocumentCollection containing the documents you want, creates a new NotesNewsletter. When used with OLE automation, this method enables you to create a NotesNewsletter object without using New.
CreateTimer	Creates a NotesTimer object. When used with OLE automation, this method enables you to create a NotesTimer object without using New.
FreeTimeSearch	Searches for free time slots for setting up a calendar and scheduling
GetDatabase	Creates a NotesDatabase object that represents the database located at the server and filename you specify and opens the database, if possible. When used with OLE automation, this method enables you to create a NotesDatabase object without using New—but does not create a new database on disk.
GetDbDirectory	Creates a new NotesDbDirectory object using the name of the server you want to access. When used with OLE automation, this method enables you to create a NotesDbDirectory object without using New.
GetEnvironmentString	Given the name of a string environment variable, retrieves its value

Methods	
GetEnvironmentValue	Given the name of a numeric environment variable, retrieves its value
New	To access the current session, use New
SetEnvironmentVar	Sets the value of a string or numeric environment variable
UpdateProcessedDoc	Marks a document as processed by an agent

NotesTimer Class

Represents a mechanism for triggering an event every fixed number of seconds.

Containment

- Contained by NotesSession
- Table 11.19 shows more information.

Table 11.19   NotesTimer

Properties	
Comment	Read-write. Any comment
Enabled	Read-write. Indicates whether the NotesTimer object is active
Interval	Read-write. The interval in seconds at which the handler for the Alarm event is called
Method	
New	To create a new NotesTimer object, use New.
Event	
Alarm	Occurs for an enabled NotesTimer object at the specified interval

NotesUIDatabase Class

Represents the database that is currently open in the Notes workspace.

Containment

- Contains NotesDatabase, NotesDocumentCollection.
- Table 11.20 shows more information.

Table 11.20 NotesUIDatabase

Properties	
Database	Read-only. The back-end database that corresponds to the currently open database
Documents	Read-only. All the documents on which the current NotesUIDatabase event is working
Method	
OpenView	Opens a view in the database
Events	
PostDocumentDelete	Occurs just after a document is deleted (cleared or cut)
PostOpen	Occurs after the NotesUIView QueryOpen and NotesIUView PostOpen events are triggered
QueryClose	Occurs just before the database closes
QueryDocumentDelete	Occurs just before a document or selected set of documents is deleted (cleared or cut)
QueryDocumentUndelete	Occurs just before a document or selected set of documents is undeleted

NotesUIDocument Class

Represents the document that is currently open in the Notes workspace.

Containment

- Contained by NotesUIWorkspace. Contains: NotesDocument.
- Table 11.21 shows more information.

Table 11.21   NotesUIDocument

Properties	
AutoReload	Read-Write. Indicates whether the current document should be refreshed whenever the corresponding back-end document changes. Refreshing the current document updates its representation in memory, and visually on the workspace, to reflect changes that have been made to the back-end document.
CurrentField	Read-only. The name of the field in which the cursor lies
Document	Read-only. The back-end document that corresponds to the currently open document
EditMode	Read-Write. Indicates whether a document is in Edit mode
FieldHelp	Read-Write. Indicates whether field help for a document is displayed
HiddenChars	Read-Write. Indicates whether the hidden characters in a document (such as tabs and carriage returns) are displayed
HorzScrollBar	Read-Write. Indicates whether a document's horizontal scroll bar is displayed
InPreviewPane	Read-only. Indicates whether the document is being accessed from the preview pane

continues

Table 11.21 Continued

Properties	
IsNewDoc	Read-only. Indicates whether a document is new. A new document is one that has not been saved.
PreviewDocLink	Read-Write. Indicates whether a link's preview pane is displayed
PreviewParentDoc	Read-Write. Indicates whether the parent document's preview pane is displayed
Ruler	Read-Write. Indicates whether the ruler is displayed
WindowTitle	Read-only. The window title of a document

Methods	
Categorize	Given the name of a category, places a document in the category
Clear	Deletes the current selection from a document. The current selection can be anything in an editable field, such as text or graphics.
Close	Closes a document. The NotesUIDocument is no longer available once you call this method.
CollapseAllSections	Collapses all the sections in a document
Copy	Copies the current selection in a document to the Clipboard. The current selection can be anything on the document, such as text or graphics.
CreateObject[1]	In a document in Edit mode, creates an OLE object in the current rich text field
Cut	Cuts the current selection from a document and places it on the Clipboard. The current selection can be anything in an editable field, such as text or graphics.

Methods

DeleteDocument	Marks the current document for deletion and closes it. The NotesUIDocument object is no longer available once you call this method.
DeselectAll	Deselects any selections in a document
ExpandAllSections	Expands all the sections in a document
FieldAppendText	Appends a text value into a field on a document without removing the existing contents of the field
FieldClear	Clears the contents of a field on a document
FieldContains	In an open document, checks whether a field contains a specific text value
FieldGetText	In a document in Read or Edit mode, returns the contents of a field you specify as a string. Whether the field is of type numbers or date-time, its contents are converted to a string.
FieldSetText	Sets the value of a field on a document. The existing contents of the field, if any, are written over.
Forward	Creates a new mail memo with the contents of a document. The user can enter recipients and mail the forwarded document, just as they can any other mail memo.
GetObject[2]	Given a name, returns a handle to the OLE object of that name
GotoBottom	Places the cursor in the last editable field or the last button on a document
GotoField	Given a field name, puts the cursor in the specified field on a document

continues

Table 11.21 Continued

Methods	
GotoNextField	Places the cursor in the next field on a document. The next field is the one below and to the right of the current field.
GotoPrevField	Places the cursor in the previous field on a document. The previous field is the one above and to the left of the current field.
GotoTop	Places the cursor in the first editable field or the first button on a document
InsertText	Inserts a text value at the current cursor position on a document
Paste	Pastes the contents of the Clipboard at the current cursor position on a document
Print	Prints the current document ■ If one or more parameters are specified, automatically prints the document ■ If no parameters are specified, or if the first parameter is omitted, displays the File Print dialog box
Refresh	Refreshes a document. When you refresh a document, its computed fields are recalculated.
RefreshHideFormulas	Recalculates the hide-when formulas on the current document's form
Reload	Refreshes the current document with any changes that have been made to the corresponding back-end document. Refreshing the current document updates its representation in memory, and visually on the workspace, to reflect changes that have been made to the back-end document.

Methods	
Save	Saves a document
SaveNewVersion	Saves a copy of a document as a new version
SelectAll	In a document in Edit mode, selects the entire contents of the current field. In a document in Read mode, selects the entire contents of the document
Send	Mails a document
Events	
PostModeChange	Occurs after the document has changed modes (from Read to Edit mode, or from Edit to Read mode), but before the user has been given input focus
PostOpen	Occurs after the current document has been opened, but before the user has been given input focus
PostRecalc	Occurs just after the current document is recalculated (after all the formulas on the document's form have executed)
QueryClose	Occurs just before the current document is closed
QueryModeChange	Occurs just before the current document changes modes (from Read to Edit mode, or from Edit to Read mode)
QueryOpen	Occurs just before the current document is opened
QuerySave	Occurs just before the current document is saved

[1]CreateObject is not supported under OS/2, UNIX, and on the Macintosh.
[2]GetObject is not supported under OS/2, UNIX, and on the Macintosh.

NotesUIView Class

Represents the current view in the Notes workspace.

Containment

- Contains NotesDocumentCollection, NotesView. Table 11.22 shows more information.

Table 11.22   NotesUIView

Properties	
CalendarDateTime	Read-only. The date and time of the current region in a calendar view
Documents	Read-only. All the documents on which the current NotesUIView event is working
View	Read-only. The back-end view that corresponds to the currently open view
Events	
PostDragDrop	Occurs just after dragdrop operation in a calendar view
PostOpen	Occurs just after the current view is opened
PostPaste	Occurs just after a document is pasted into the view
QueryAddToFolder	Occurs just before an add-to-folder operation
QueryClose	Occurs just before the view is closed
QueryDragDrop	Occurs just before a dragdrop operation in a calendar view
QueryOpen	Occurs just before the current view is opened
QueryOpenDocument	Occurs just before a document is opened from this view

Events	
QueryPaste	Occurs just before a document is pasted into the view
QueryRecalc	Occurs just before the view is recalculated
RegionDoubleClick	Occurs just after the current region is double-clicked in a calendar view

NotesUIWorkspace Class

Represents the current Notes workspace window.

Containment

- Contains NotesUIDocument. Table 11.23 shows more information.

Table 11.23   NotesUIWorkspace

Properties	
CurrentCalendarDateTime	Read-only. The date and time of the current (selected) region in a calendar view
CurrentDocument	Returns a NotesUIDocument object representing the document that is currently open
Methods	
AddDatabase	Adds a database to the workspace and highlights the icon, or highlights the icon if the database is already on the workspace
CheckAlarms	Triggers the alarm daemon to check for new alarms in the mail file

continues

Table 11.23 Continued

Methods	
ComposeDocument	Using a database and a form you specify, creates a new document and displays it on the workspace
DialogBox	Brings up a dialog box that displays the current document (either open or selected in a view) or a specified document using a form you specify. The user interacts with the form and document as usual, clicking OK or Cancel when finished. This function can be used with any form, but it is particularly useful with forms containing a single layout region, because the user can interact with the layout region as if it were a dialog box.
EditDocument	Opens the current or a specified document in a mode you specify. The current document may be either ■ The document that is currently open ■ The document that is currently selected in a view or folder
EditProfile	Opens a new or existing profile document in Edit mode
EnableAlarms	Starts or stops the alarm daemon
FindFreeTimeDialog	Brings up the Find Free Time box
New	To create a new NotesUIWorkspace object, use New
OpenDatabase	Opens a database to a view you specify
URLOpen	Retrieves a World Wide Web page specified by its URL

Methods	
UseLSX	Loads a LotusScript extensions (lsx) file containing Public definitions
ViewRefresh	Refreshes the current view

NotesView Class

Represents a view or folder of a database and provides access to documents within it.

Containment

- Contained by NotesDatabase, NotesUIView. Contains: NotesDocument and NotesViewColumn. Table 2.24 shows more information.

Table 11.24   NotesView

Properties	
Aliases	Read-only. The aliases of a view
AutoUpdate	Read-write. Indicates whether the front-end view is automatically updated each time a change occurs in the back-end
Columns	Read-only. All the columns in a view
Created	Read-only. The date that a view was created
IsCalendar	Read-only. Indicates whether a view is a calendar view
IsDefaultView	Read-only. Indicates whether a view is the default view of the database
IsFolder	Read-only. Indicates whether a NotesView object represents a folder

continues

Table 11.24 Continued

Properties	
LastModified	Read-only. The date that a view was last modified
Name	Read-only. The name of a view
Parent	Read-only. The database to which a view belongs
ProtectReaders	Read-write. Protects $Readers items from being overwritten by replication
Readers	Read-write. The contents of the $Readers field associated with the view
UniversalID	Read-only. The Universal ID of a view, which is a 32-character combination of letters and numbers that uniquely identifies a view across all replicas of a database

Methods	
Clear	Clears the full-text search filtering on a view
FTSearch	Conducts a full-text search on all documents in a view and filters the view so that it represents only those documents matching the full-text query. This method does not find word variants.
GetAllDocumentsByKey	Finds documents based on their column values within a view. You create an array of strings (keys), where each key corresponds to a value in a sorted column in the view. The method returns all documents whose column values match each key in the array.

Methods

GetChild	Given a document in a view, returns the first response to the document
GetDocumentByKey	Finds a document based on its column values within a view. You create an array of strings (keys), where each key corresponds to a value in a sorted column in the view. The method returns the first document whose column values match each key in the array.
GetFirstDocument	Returns the first document in a view. This document is the same one you see when you scroll to the top of the view in the Notes user interface.
GetLastDocument	Returns the last document in a view. This document is the same one you see when you scroll to the bottom of the view in the Notes user interface.
GetNextDocument	Given a document in a view, returns the document immediately following it
GetNextSibling	Given a document in a view, returns the document immediately following the given document at the same level. If you send the method a main document, the next main document in the view is returned. If you send a response document, the next response document with the same parent is returned.
GetNthDocument	Given a number, returns the document at the given position in the top level of a view
GetParentDocument	Given a response document in a view, returns its parent document
GetPrevDocument	Given a document in a view, returns the document immediately preceding

continues

Table 11.24 Continued

Methods	
GetPrevSibling	Given a document in a view, returns the document immediately preceding the given document at the same level. If you send the method a main document, the preceding main document in the view is returned. If you send a response document, the preceding response document with the same parent is returned.
Refresh	Updates a view's contents with any changes that have occurred to the database since the NotesView object was created—or since the last Refresh
Remove	Permanently deletes a view from a database

NotesViewColumn Class

Represents a column in a view or folder.

Containment

- Contained by NotesView. Table 11.25 shows more information.

Table 11.25 NotesView Column

Properties	
Formula	Read-only. The @function formula for a column, if it exists
IsCategory	Read-only. Indicates whether a column is categorized
IsHidden	Read-only. Indicates whether a column is hidden

Properties	
IsResponse	Read-only. Indicates whether a column contains only response documents
IsSorted	Read-only. Indicates whether a column is sorted
ItemName	Read-only. The name of the item whose value is shown in the column. For columns whose values are simple functions or formulas, returns an automatically generated internal name
Position	Read-only. The position of a column in its view. Columns are numbered from left to right, starting with one.
Title	Read-only. The title of a column, if any

ODBCConnection Class

Represents the ODBC Data Access features for connecting to a data source. Table 11.26 shows more information.

Table 11.26   ODBCConnection

Properties	
DataSourceName	Read-only. The ODBC name of the connected data source
DisconnectTimeOut	Read-write. The number of seconds that a connection is maintained before timing out and disconnecting
Exclusive	Read-write. Sets a mode where a separate connection is made, rather than using an existing connection that the DBMS may have cached
IsConnected	Read-only. Indicates whether the physical connection is valid

continues

Table 11.26   Continued

Properties	
IsSupported	Read-only. Indicates whether an option is supported
IsTimedOut	Read-only. Indicates whether a connection has timed out
SilentMode	Read-write. Prevents a User ID and Password dialog box from being displayed during program execution

Methods	
ConnectTo	Connects to a data source
Disconnect	Disconnects from the data source
ExecProcedure	Executes a stored procedure
GetError	Returns an error code value
GetErrorMessage	Returns the short error message text associated with any error value
GetExtendedErrorMessage	Returns the long, descriptive error message text associated with an error code value
GetRegistrationInfo	Returns the parameter setting stored in the registration file as text
ListDataSources	Lists all registered data sources as an array
ListFields	Returns an array of field (column) names in a table
ListProcedures	Lists the executable procedures at the connected data source
ListTables	Lists the available tables in a connected data source

ODBCQuery Class

Represents the ODBC data access features for defining a SQL statement. A query must pertain to a valid connection before it can be used or validated. Table 11.27 shows more information.

Table 11.27   ODBCQuery

Properties	
Connection	Write-only. Relates a connection object with a query
QueryExecuteTimeOut	Read-write. Specifies a timeout value in seconds for query execution
SQL	Read-write. Any SQL statement
Methods	
GetError	Returns an error code value
GetErrorMessage	Returns the short error message text associated with any error value
GetExtendedErrorMessage	Returns the long, descriptive error message text associated with an error code value

ODBCResultSet Class

Represents the ODBC data access features for performing operations on a result set. Table 11.28 shows more information.

Table 11.28   ODBCResultSet

Properties	
Asynchronous	Read-write. Asynchronous mode
AutoCommit	Read-write. Indicates whether a result set operates in auto-commit mode

continues

Table 11.28 Continued

Properties	
CacheLimit	Read-write. The maximum number of rows to be cached in memory
CommitOnDisconnect	Read-write (whether transactions are committed or rolled back on a disconnect)
CurrentRow	Read-write. The number of the current row in a result set, starting from row one
FetchBatchSize	Read-write. Specifies the number of records that will be fetched at any one time. The default is one (get just the requested record).
HasRowChanged	Read-only. Indicates whether values in the current row have changed in the data source; that is, whether they differ from what was fetched into the result set
IsBeginOfData	Read-only. Indicates whether the cursor position is at the beginning of the result set
IsEndOfData	Read-only. Indicates whether the cursor position is at the end of the result set
IsResultSetAvailable	Read-only. After a query, indicates whether the query yielded any data
MaxRows	Read-Write. Maximum number of rows fetched to the result set, or zero to fetch all rows subject to memory limits
NumColumns	Read-only. Number of columns in a result set
NumRows	Read-only. Number of rows in a result set

Properties	
Override	Write-only. Overrides another user's changes to the data source in response to a RowContentsChanged event after an UpdateRow operation
Query	Write-only. Relates an ODBCQuery object with a result set
ReadOnly	Read-write. Indicates whether a result set is read-only. Can be set to True to prevent updates
Methods	
AddRow	Adds a workspace for a new row
Close	Closes a result set and commits or rolls-back changes to the external data source
DeleteRow	Deletes the record represented by the current row of the result set from the specified table in the external database
Execute	Executes an SQL statement
FieldExpectedDataType	Gets or sets the data type for a column
FieldID	Returns the field ID of a field as an integer value, given the field name
FieldInfo	Returns an array of field information
FieldName	Returns the field name for a specified column, given its ID (column number)
FieldNativeDataType	Given a specific column, returns the native data type for that column
FieldSize	Returns the maximum size of data for a field

continues

Table 11.28 Continued

Methods	
FirstRow	Sets the current row to the first row in a result set. Equivalent to setting the CurrentRow to one
GetError	Returns an error code value
GetErrorMessage	Returns the short error message text associated with any error value
GetExtendedErrorMessage	Returns the long, descriptive error message text associated with an error code value
GetParameter	Returns the last value set for the specified parameter
GetParameterName	Retrieves a parameter name
GetRowStatus	Returns the status of the current row
GetValue	Retrieves the value of the column referenced by *column_id%* and copies the results to the *variable* field
IsValueAltered	Indicates whether a column value was altered by SetValue
IsValueNull	Indicates whether a column value is a null data value
LastRow	Sets the current row position to the last row in the result set
LocateRow	Identifies records that match a specified field value
NextRow	Sets the current row position to the next row in a result set
NumParameters	Returns the number of parameters in an SQL statement
PrevRow	Positions the current row pointer at the previous row

Methods	
RefreshRow	Refreshes from the data source the data cached at the row pointed to by the current row pointer
SetParameter	Sets the replacement value for the specified parameter
SetValue	Sets a new value to a column
Transactions	Commits or rolls-back changes in a data source that uses the transaction model
UpdateRow	Updates the current row, if changed, to the database

Index

About the Author

William Thompson, CLI, CLP, and CLS, consults for Whitman-Hart, Inc. He teaches LotusScript at Whitman-Hart as well as to clients. He began his IS career at Anderson Consulting, where he developed applications using Lotus Notes.